INDIAN MIGRATION AND EMPIRE

INDIAN MIGRATION AND EMPIRE

A Colonial Genealogy of the Modern State

RADHIKA MONGIA

DUKE UNIVERSITY PRESS *Durham and London* 2018

Typeset in Minion Pro by Copperline Book Services.

Library of Congress Cataloging-in-Publication Data
Names: Mongia, Radhika, [date–] author.
Title: Indian migration and empire : a colonial genealogy
of the modern state / Radhika Mongia.
Description: Durham : Duke University Press, 2018. |
Includes bibliographical references and index.
Identifiers: LCCN 2017058619 (print) |
LCCN 2018000889 (ebook)
ISBN 9780822372110 (ebook)
ISBN 9780822370390 (hardcover : alk. paper)
ISBN 9780822371021 (pbk. : alk. paper)
Subjects: LCSH: India—Emigration and immigration—
Social aspects—History. | India—Emigration and
immigration—Government policy—History. | Contract
system (Labor)—India—History. | Great Britain—
Colonies—Emigration and immigration. | Passports—
Government policy—India—History.
Classification: LCC JV6225 (ebook) | LCC JV6225 .M67 2018
(print) | DDC 325/.273—dc23
LC record available at https://lccn.loc.gov/2017058619

Cover art: Reena Kallat, *Anatomy of distance*, 2013.
Courtesy of Reena Kallat Studio.

IN MEMORY OF MY PARENTS,

VED VYAS MONGIA (1927–1988) AND

USHA MONGIA (1932–2017)

CONTENTS

ACKNOWLEDGMENTS

For personal reasons, this book needed to be put aside for many years. But in those years it was, of course, never out of my mind. I am happy to have had the time to let the ideas presented here percolate and it is my hope that the resulting brew is better for the time it has taken.

Over the years, several people have either commented on all or portions of this project or have discussed issues addressed here. For their insights, engagements, and provocations, I thank Itty Abraham, Antoinette Burton, Dipesh Chakrabarty, Partha Chatterjee, Dilip Gaonkar, Lawrence Grossberg, Kajri Jain, Samantha King, Stefan Kipfer, Renisa Mawani, Philip McMichael, Melissa Orlie, David Prochaska, Anupama Rao, Vanita Seth, Nandita Sharma, Mrinalini Sinha, Carol Stabile, Rachel Sturman, and Dale Tomich. Most especially, I thank Aparna Sundar, for reading the penultimate draft of the entire manuscript in short order, and the anonymous reviewers for Duke University Press, for their enthusiastic yet challenging comments.

The long time it has taken to send this book to press has meant that the people who have contributed to its making have grown in equal measure. For conversations and confabulations that informed early stages of this project, I thank Jack Bratich, the late Greg Dimitriadis, Suvir Kaul, Lisa King, Samantha King, Ania Loomba, Carrie Rentschler, Craig Robertson, Jonathan Sterne, and Daniel Vukovich. Many family, friends, and colleagues have been the source of support, humor, gossip, merriment, distractions, engagements, encouragement, threats, and solace. For thus enlivening my life and my thinking, perhaps unbeknown to them, I thank Ajeya Raj Adhikari, Mira Adhikari, Vyas Adhikari, Mark Anderson, Nisha Agrawal, Paola Bacchetta, Himani Bannerji, Nigel Barriffe, Atiya Bose, Shonali Bose, Deborah Brock, Sylwia Chrostowska, Ena Dua, Lorna Erwin, Andil Gosine, Jennie Haw, Rosemary Hennessy, Gail Hershatter, Emily Honig, James Ingram, Maki Iwase, Toshio Iyotani, Alok Johri, Miranda Joseph, Rohini Kandhari, Ilan Kapoor, Kamala Kempadoo, Terry Maccagno, Gayatri Menon, Amira Mittermaier, Kent Murnaghan, Eric Mykhalovskiy, Naintara Maya Oberoi, Naira Oberoi, Vinay Sheel Oberoi, V. Spike Peterson, Jim Phillips, Bhavani Raman, Shilpa Ranade, Dereka Rushbrook, Nayan Shah, Nicola Short, Anil Singh, Ari Singh, Rupinder Singh, Sandy Soto, Deborah Sutton, Megan Thomas, Alissa Trotz, Joanna Van Gruisen, and Cynthia Wright.

The project entailed much travel and I am grateful to many people. In South Africa, the hospitality and generosity of Martin Gustafsson, Ilva MacKay Langa, Melissa Mackay, Naomi MacKay, and, most especially, the marvelous Shereen Mills, facilitated my trips to the archives and provided a wonderful network of friends and interlocutors in Johannesburg and Pretoria. I thank the Indian Consulate in Durban, for facilitating trips to the archives in Pietermaritzburg, and Ashwin Desai, for finding time to hang out in Durban. In Mauritius, I am very grateful for the hospitality of the Ramdeen family and the Indian Consulate. In Delhi, I have been invigorated by conversations, over the years, with Uma Chakravarty, Charu Gupta, Riyad Koya, Prabhu Mohapatra, and Radhika Singha.

The interlocutors at numerous invited talks at various venues in Canada, Germany, India, Japan, the United Kingdom, and the United States have enriched and greatly enhanced this book. A Rockefeller Residency Fellowship at the University of Arizona, Tucson, and travel research grants from the Committee on Research at the University of California, Santa Cruz, and from the Faculty of Liberal Arts and Professional Studies at York University, Toronto, were indispensible to this project. I thank Rukun Advani and, especially, Ken Wissoker, at Permanent Black and Duke, respectively, for their boundless patience and understanding.

Portions of chapters 1 and 4 appear, in different form, in *Comparative Studies in Society and History* 49, no. 2 (2007). Aspects of chapter 3 also appear in *Gender and History* 18, no. 1 (2006). An earlier version of chapter 4 was published in *Public Culture* 11, no. 3 (1999).

My life would be much impoverished and my ability to work greatly hampered without a web of deep and enduring friendships. I am immensely grateful to the many friends who have bid me farewell and welcomed me back, time and again, with compassion and joviality, as I have shuttled endlessly between India and Canada/North America. I thank Shailaja Bajpai, Viraj Chopra, Shubhra Gururani, Kajri Jain, Samantha King, Srirupa Roy,

Marya Ryan, Vanita Seth, Aparna Sundar, Lalit Vachani, and Anna Zalik for seeing me through a rocky decade of departures and arrivals and for endless conversations, intellectual and otherwise. It would take too many words to adequately express my good fortune in the sisters I have, who have been the source of abundant care, unconditional love, and infinite kindness: Nandini Oberoi, Padmini Mongia, and Sonia Mongia Adhikari keep me warm and safe in an embrace that stretches countries and continents. To them I owe more than they can know. Sadly, my mother, Usha Mongia, passed away just months before this book went to press, but not before she had shown me the meaning of sweetness, of an expansive love, and of enormous patience and good humor, even in the face of a brutal and protracted illness. I dedicate this book to her, model patient and incomparable mother, and to my father, Ved Vyas Mongia. Both desperately missed.

The global monopoly of a system of states over the international movement of people seems an unremarkable fact in the present world. *Indian Migration and Empire: A Colonial Genealogy of the Modern State* analyzes how this came to be the case. In overarching terms, two questions frame this study: First, what histories can we chart of the increasing and incremental state control over migration that culminate, by the early decades of the twentieth century, in a state monopoly over migration? Second, what can these histories tell us about state formation, inter-state relations, state sovereignty, and modern subject constitution? I argue that since the state has not always held a monopoly over migration, investigating the historical circumstances and the historical human subjects that propelled it to do so yield crucial insights into the processes of modern state formation. In particular, as this work will delineate, such an investigation forces a reevaluation of the common distinction between the metropolitan/modern state and the colonial state and, instead, impels an approach that traces what I will call a *colonial genealogy of the modern state.*

Focused on an analysis of colonial Indian migration from the British Empire, this study traces a shift from a world dominated by empire-states into a world dominated by nation-states.[1] I begin in approximately 1834, with the abolition of plantation slavery in British colonies, which generated an acute demand for labor in the labor-intensive plantation economies.[2] This demand was met, in part, via the introduction of laborers from India. As early as 1835, the Court of Directors of the East India Company (in charge, at the time, of British administration in India) installed mechanisms of state control to monitor this massive movement.[3] As such regulation lacked legal precedence, it occurred amid challenges to the authority and legality of the state in monitoring the movement of "free" subjects (as distinct from the "unfree" subjects of the African and other slave diasporas). While recognizing that "this practice [had] no foundation in any existing law," the state argued that the regulations were warranted in order to ensure the migration was "free."[4] Thus, this early state involvement in colonial Indian migration saw its genesis in facilitating and enabling the movement. Beginning with a trickle of thirty-nine Indian migrants who arrived in Mauritius in 1834, such state-supervised migration grew into a vast enterprise operating within a *logic of facilitation*.[5] Under its aegis, more than 1.3 million Indians moved to various parts of the globe, including Mauritius, Réunion, Guyana, Trinidad, Jamaica, Surinam, Fiji, Australia, Tanzania, Kenya, Uganda, and South Africa.

However, until the late nineteenth and early twentieth centuries, the state monitored only the movement of indentured Indian labor and did not interfere with the much larger migration of those not participating in the state-defined and state-controlled indenture system.[6] Within the history of colonial Indian migration, it is only with the increasing movement of non-indentured Indians to white-settler colonies like Canada, Australia, South Africa, and the United States, in the first two decades of the twentieth century, that we see persistent demands to extend state control to cover all types of migration to prohibit and restrict movement. These twentieth-century demands, and their eventual incorporation into a global regime of migration control, operated on a *logic of constraint*. In this study, I consider both the predominantly nineteenth-century migration of indentured Indian labor to replace—or, more accurately, to displace—recently emancipated slave labor in the British plantation economies, as well as the early twentieth-century movement of non-indentured migrants to white-settler colonies (specifically, Canada and South Africa).[7] I focus on the emergence of certain techniques, technologies, and institutions for managing migration: the juridical labor contract, installed to distinguish indentured migration from the slave trade; the formation of a complex and gargantuan bureaucracy that deployed a range of management techniques for monitoring migration; the marriage license, required of "free," non-indentured women to stake a claim to mobility; and the modern passport, instituted to control migration along the axis of racialized nationality. This book charts how these technologies and institutional forms emerged from a circuit of connections and contestations between imperial and colonial state formations; the expansion and segmentation of global regimes of capital; prevailing ideologies of race, gender, and sexuality; competing nationalisms; and complex reconfigurations in the meanings of "labor," "freedom," "family," and "marriage" in places as diverse as India, Britain, Mauritius, the Caribbean, Canada, and South Africa.

The debates attendant on the historical formation of these technologies were centrally concerned with such issues as the proper and legitimate purview of the state, the status of "free" subjects, the definition of state sovereignty, and the nature of inter-state relations. Indeed, my analysis demonstrates that the formation of colonial migration regulations was dependent upon, accompanied by, and generative of profound changes in normative understandings of the modern state. The analysis points, in other words, to a fundamental colonial genealogy of the modern (nation-) state, in *both* the metropoles and the colonies, or *globally*. It thus departs from what one can call a dispersal or diffusionist model of modern state formation. Unlike the claims of this model, wherein the modern (nation-) state takes shape in Euro-America and, over time, spreads outward, this book argues that the modern (nation-)state has a far more complex and complicated history, one whose coordinates and determinations are temporally and spatially dispersed.

The central chapters here are organized to illuminate the formation of key techniques and technologies for regulating migration. Studying migration by employing state technologies of regulation as the object of analysis allows one to consider patterns of migration often held distinct (for instance, indentured and "free" migration) within the rubric of one project. This approach also allows one to address the problems of methodological nationalism, which sees the national as the privileged site and scale for investigating migration and, thereby, misunderstands how definitions of the "national" are necessarily implicated in, and emerge from, non-national, cross-statal, transcolonial, and inter- and intra-imperial forces. Scholarship that considers different patterns of migration does so largely within the rubric of a comparative framework. My aim here is not to foreground a comparative analysis; it is, rather, to map the unique historical exigencies that propel certain modes of state control over migration, which come to be congealed in certain, significantly enduring, technologies. These technologies, I show, both embody and express critical moments in the making of—and transformations in—the modern state. With the larger objective of producing what I am calling a colonial genealogy of the modern state, this study illuminates significant aspects of colonial Indian migration, examines the place of migration in the transformation of a world dominated by empire-states into a world dominated by nation-states, and explores how local particularities are both encoded in and help catalyze the development of global regularities in the consolidation of state control over mobility.

In the remainder of this introduction, I review some paradigms that shape migration research in relation to the state, outline what I mean by a "colonial genealogy of the modern state," and, finally, provide an overview of the chapters that follow.

Migration/Nation/State: Analytical Paradigms

As with the human sciences more broadly, the analysis of migration has been framed (in terms both of providing an *orientation* for analysis and of serving to *enclose* and delimit the analysis) by a profound methodological nationalism. The framework of methodological nationalism works on a foundationalist and presentist logic, wherein the nation-state is divested of its historicity and peculiarity and invested, instead, with a dehistoricized and circular logic. In this way, the very notions—of nation, nationality, and the nation-state—most in need of explanation are both the starting and the end point of analysis. Thus, for instance, migration was customarily understood as the movement of people from one nation-state to another, and analyses, working with the paradigm popularized by the Chicago School, were largely configured to inquire into how migrants were assimilated, or not, into a "host" society understood as a preexisting national space. A corollary to the assimilationist paradigm, focused on assessing migration as a phenomenon of *im*migration, had been the lack of attention to understanding migration as, simultaneously and necessarily, also *e*migration.[8] In recent decades, the implicit and explicit methodological nationalism that has dominated the study of migration has received sustained critiques from at least two quarters. One critique, associated with the influential work of cultural studies, has vigorously questioned notions of a fixed, stable culture or cultural identity, in general, and notions of unchanging, discrete national culture, in particular.[9] Another critique has emphasized how the insular, territorial logic of methodological nationalism obstructs our ability to grasp the dense networks and circuits of connections between people, discourses, commodities, artifacts, ideas, and so forth, that extend beyond national-state borders. A "transnational" approach has been one of the most significant innovations in addressing these limitations.[10] Both critiques are important interventions that have significantly shaped recent migration research. While drawing on their insights, I aim here for a more radical historicization of the state as a territorially and demographically contained entity than is evident in these approaches. This task entails that we foreground the historicity of the regulatory categories that organize migration regimes and thus ensure that our "categories of analysis" do not uncritically duplicate—and thereby naturalize—the "categories of practice" deployed by the state.[11] In fact, what we can call a *methodological statism*—a position that naturalizes the state—is embedded in both cultural studies and transnational migration studies and, despite some fissures, remains pervasive in migration scholarship more broadly. With regard to cultural studies scholarship, analyses of the state have been largely absent, even when certain (national) state spaces are the central locus of concern. Though we have rich accounts of the (re)making of national cultures, a methodological statism enters by way of an assumption of a dehistoricized, invariant, usually coercive state that frames and orients the analysis but is not itself subjected to examination. In the transnational approach, the problems of a methodological statism are more acute, if paradoxical. Given that its main objective is to inquire into processes and practices that cross the boundaries of the nation-state, the existence of the nation-state and the salience of particular understandings of borders between states are central to definitions of transnationalism and transnational migration.[12] By distinguishing between state space and social space, and focusing attention on the latter, what is lost is a historicization of how, when, and why borders between malleable states come to be congealed or of what events and processes produce borders as containing fixed territories and populations understood in specifically national terms.[13] While a transnational approach is useful for understanding certain recent (trans)formations, it is not as helpful for historical inquiry since, in its reliance on the "national," it introduces the problem of presentism in a particularly acute yet unacknowledged fashion. The problem is embedded in the very nomenclature: the formulation of the trans*national* obliges if not shackles us to assumptions of space, state, and subjectivity *already* conceived in *national* terms.[14]

A chief reason for the confusing deployment of the national framework in migration scholarship relates to the fact that mass migrations have not been an important element of influential treatments of the historical development of nationalism and the nation form.[15] Rather than *assume* that a state monopoly over migration and a state organization of migration in national terms were teleologies simply waiting to unfold, we must examine *how* certain events in certain historical conjunctures produced a tight confluence between migration, nationness, and stateness as a contingent—if enduring—result. Rogers Brubaker's distinction between developmentalist and eventful perspectives in thinking about nationhood proves useful here. Whereas the developmentalist literature "traces long-term political, economic, and cultural changes that led, over centuries, to the gradual emergence of nations," an eventful perspective thinks of "nationness as an event, as something that suddenly crystallizes rather than gradually develops, as a contingent, conjuncturally fluctuating, and precarious frame of vision and basis for individual and collective action, rather than a relatively stable product of deep developmental trends in economy, polity, or culture."[16]

The historical nationalization of migration is best grasped through an eventful perspective, where certain migrations, at particular moments, come suddenly to provoke the framing of identity in national and nationalist terms, or to catalyze the introduction of nationality as an institutionalized category into migration law, or to produce unforeseen eruptions of fervent nationalist claims. Historically, as we will see, the nationalization of migration has taken a piecemeal and uneven trajectory, pointing to the fact that processes of nationalization are "temporally heterogeneous."[17] These processes do not all work in tandem, do not all have the same intensity, and nationalization in one domain or in some state-territorial spaces does not entail, or foretell, nationalization in others. Given this temporal–spatial heterogeneity, while we might now speak of a thorough nationalization of migration on a global scale, the particularities of this nationalization do not all replicate each other and are unstable. Moreover, as I will demonstrate in chapter 3, an eventful approach permits an analysis of how gendered and sexualized determinations shape the nationalization of migration, an analysis that largely cannot be, and has not been, accommodated within developmentalist approaches.

Another important aspect of methodological statism that characterizes scholarship on migration is the assumption that controlling migration across putative state borders is a long-standing and noncontentious element of state sovereignty. Scholars of migration have noted that different kinds of states, at different times, have sought to control the mobility of people, including what we call emigration and immigration, and that the axes for such control have varied widely.[18] They have also noted that several states, such as the United States, that have seen a flow of migrants for centuries, attempted to develop comprehensive federal immigration laws only toward the end of the nineteenth century.[19] Both past and current migration policies are increasingly scrutinized and debated. Scholars have also understood the extent and mechanisms of control, particularly on emigration, as yardsticks to classify state forms as, for instance, liberal or totalitarian.[20] In sum, several elements of state control over mobility have received attention. However, even as these different aspects have occupied migration scholars, they have tended, with some exceptions, to assume that such control is a defining, definitive, unchanging, and unchangeable element of (state) sovereignty, typically attributing this feature to the 1648 Treaty of Westphalia.[21] In other words, the different policies, legislative actions, and mechanisms of control—over almost four centuries—are, if implicitly, understood as so many diverse "applications" of the "principles" or doctrines implied by the Treaty of Westphalia. As a result, the "nationalization" of sovereignty—in terms of how preexisting aspects of other state forms are remade in national terms and of how new, uniquely national, forms emerge—are only recently receiving attention.[22] Such appeals to a Westphalian ideal of the state and the interstate system not only rely on an insufficient historicization of the state and state sovereignty in relation to migration. They also sidestep the issue of how—and if—the Westphalian ideal was globalized and if we might chart any relations between this ideal and colonialism.[23] More broadly, such a view is premised on the notion that the practices of governance and the institutions of the state have a fidelity to, can be deduced from, and are reflective of a set of principles or of treaties. In other words, this view conflates what Philip Abrams calls the "state-system" and the "state-idea."[24] Rather than understand the state as a coherent, interpretatively stable set of principles that are put into practice and encoded in institutions, I focus on practices, techniques, and institutions to examine how they come into being and how they encode and remake principles *in particular historical conjunctures*. Thus, the state will emerge, in my analysis, as an unstable, historically changing entity, rather than as an entity that adheres to principles and fulfills static, definitional criteria.[25] Such a perspective will enable us to avoid the pitfalls of a methodological nationalism that would suggest that migration has always been controlled in national terms and that state sovereignty embodies an inviolable right to exercise such control.

For instance, I will demonstrate in chapter 1 that, notwithstanding the 1648 Treaty of Westphalia, in the nineteenth century the British empire-state was beset with problems in legitimizing state control over Indian indentured migration that followed the abolition of slavery.[26] As we consider the debates, contentions, and legal conundrums attendant on legitimizing such control, which institutionalized migration under a state-authorized labor contract, we are forced to question the methodological statism within migration scholarship that has so readily adopted the state's own *current* claims about its inviolable authority in regulating migration.[27] Such an unequivocal principle, which would have substantially eased the legal difficulties of the British empire-state in regulating—albeit to facilitate—"free" Indian migration in the nineteenth century, was, however, unavailable at the time. Thus, the legitimization of control came not by way of an appeal to a principle that authorized such control, but by way of an appeal to the *exceptional* nature (that is, the civilizational deficit) of colonized subjects that warranted a *deviation* from the principle, or the prevailing norm, that did *not* authorize such control. Chapter 4 further develops the historical contours of transformations in understandings of state sovereignty and its relation to migration with an analysis of the debates, some seventy years later in the early twentieth century, regarding Indian migration to Canada. I demonstrate how prohibiting Indian migration was achieved through a new and novel understanding of *national* sovereignty incorporating the "principle" of "reciprocity." In each instance, the shape of state control was justified via ad hoc arguments that spoke to the specificities of the precise historical conjuncture that obtained; such ad hoc resolutions to historically contingent circumstances would, over time and with repetition, come to be standardized and make anew the naturalized understandings of the relation between state sovereignty and migration control. It was not that a stable understanding of state sovereignty—a doctrine or a set of principles emanating from Westphalia and Europe—was simply "applied" or "dispersed" to the empirical "cases" at hand; rather, the matters at hand and the specificity of the empirical situation would catalyze debate, come to be embedded in the doctrine, and remake the "theory." To challenge the dispersal model of state (trans)formation that informs dominant trends in migration studies and enable us to apprehend the transactions—between principles/doctrine and empirical circumstances, between "theory" and "history," between Europe and its Others—this study adopts frameworks more attuned to coproduction and co-constitution.

We have naturalized the state at smaller and more mundane scales as well. It is, in fact, commonplace to study human movement in terms of state categories of regulation and legal distinction. When considered in a historical frame, perhaps the most important distinction is that between "free" and "unfree" movements. As with the division of migrants by nationality, the free/unfree distinction is embedded in migration scholarship. David Eltis points out that analyses "have not only been overwhelmingly country-specific, they have also focused exclusively on either free or coerced streams of migrants, at least insofar as transoceanic movements of people are concerned."[28] Despite important exceptions, migration scholarship posits a demarcation between "free" and "unfree" migration—treating them as incomparable phenomena and studying them in isolation from each other.[29] Not only are free and coerced migrations thought to be unrelated; studying them in isolation from each other also produces a reified understanding of "freedom" as equivalent to consent embodied in a contract—the very understanding of "freedom" propagated by the state conjoined with discourses of political economy.[30] This assumption has largely restricted analysis of state regulation firmly within the sphere of "free" migration. As a result, despite the involvement of states in managing, facilitating, and, eventually, rendering illegal "unfree" and "semi-free" movements, such moments of state regulation are excised from histories and theories of state regulation of migration that have come to focus only on migrations termed "free."[31]

Given this assumption, historians and theorists of migration are generally agreed that widespread state control of migration is a distinctly twentieth-century phenomenon.[32] Two further presumptions ground this assessment: first, that the "origins" of migration regulations lie in Western Europe and the United States (and, to a lesser extent, in other white-settler colonies such as Australia and South Africa) and, second, that policies of restriction and prohibition are the defining elements of migration regulation; in other words, that states regulate migration primarily within a logic of constraint. These presumptions inform dominant paradigms of migration theory, which rarely consider patterns of migration in other regions as the empirical material for formulating larger theoretical frameworks.[33] Indeed, the link between Euro-America and the focus on "free" migration is now so deeply lodged in migration theory that though historians working, for instance, on indentured Indian migration have repeatedly noted that it was a state-regulated and state-managed system, the implications of this distinctive feature have received little theoretical attention in terms of political theories of state formation.[34] Thus, while historians of colonial migrations have provided detailed descriptions of the manifold migration controls undertaken by a variety of states, they have not assessed these controls as profoundly intertwined with the historical development of, and transformations in, the modern state and inter-state relations.[35]

While the free/unfree distinction is central to historical investigations of migration, as state categories for distinguishing forms of human movement have mutated and proliferated, the foci of migration research has moved in tandem, leading to studies organized according to these new categories. Like the state, research routinely distinguishes "political refugees" from "economic migrants," "legal" migrants from "illegal(ized)" migrants, "highly skilled workers" from "guest workers," "trafficked victims" from "international terrorists," and so on, often with less than salutary effects.[36] Dwelling on this trend, Diana Wong interrogates the troubling proximity, indeed replication, between the categories that frame the migration *policy* agenda of Western "receiving countries" and the categories that have come to frame the migration *research* agenda of scholars.[37] Or, to use Brubaker's formulation, there is an overlap between categories of practice and categories of analysis. Wong argues there is a "Northern bias" in migration research, which is replicated within research on and from the South; that, in fact, migration analysis is overwhelmingly framed from the perspective of Western "receiving countries."[38] Thus, migration scholarship—though emerging from and concerned with different locales—adopts not the position of just any state; it has tended to adopt the position of certain states, revealing an enduring (neo)Eurocentrism.[39] Rather than understand modalities of state control over mobility as either peripheral to or a mere reflection and logical consequence of certain state formations, this study approaches the regulation of migration as an important site of state (trans) formation, globally.

A Colonial Genealogy of the Modern State

The last two decades have seen intensified discussion and analysis of the colonial state and its explicit or implicit other, the metropolitan state, most often simply called the modern state. Regardless of the locale or the kind of colonial formation under study, there is wide agreement that a distinguishing feature of the colonial state is the persistent recourse to instituting and institutionalizing differentiated legal regimes and political subjectivities.[40] If, over the course of the nineteenth and early twentieth centuries, the polity in the metropole would, in different measure and with different degrees of success, come to be seen as constituted by formally equal and equivalent citizens, the polity in the colony was riven with multiple stratified distinctions between the colonizer and the colonized.[41] Depending on the particular historical conjuncture, the justificatory logic for such differentiation varied; however, one can discern at least two broad patterns with regard to the British Empire. In one iteration, despite an overarching universalist liberal ideology, the differentiations between metropole and colony and within the colonial world were defended on the grounds of the exigencies produced by the peculiar situations of the colonies and the practices of the natives. Partha Chatterjee has called this operation—wherein the colonial situation necessitates an exception to (otherwise) universally valid principles—"the rule of colonial difference."[42] Frequently, these exceptions were conceived as temporary suspensions in a teleological, social-evolutionary schema that, over time, would be eliminated.[43] In another iteration, universalist and universalizing claims recede, with the cultural difference of the colony itself justifying the need for unique principles and modes of governance to preserve and protect those differences.[44] But "difference" here is not conceived as incomparable, pure alterity; rather, the differences are nonetheless brought within the ambit of a comparative—and hence normative and hierarchical—Eurocentric framework.[45] Despite the differing rationales, in both iterations of colonial rule, which largely operated in conjunction, not only was the colony governed in ways distinct from the metropole; additionally, metropolitan subjects in the colony were treated differentially.[46] Thus, scholars have explored how practices and principles "imported," "imposed," or "diffused" from the metropolis would take a very different form and shape in the colonies, forms that have variously been characterized as "exception," "distortion," "incompleteness," or "hybridity."

However, a focus on the distinctiveness of the colonial state or on the fate of metropolitan principles and practices in the colonies leaves important aspects of both the colonial and the metropolitan state in shadow. The assumption of spatial–territorial segregation that subtends the distinction between the colonial state and the metropolitan state disallows an investigation into how "modern" forms and institutions of state had a developed life at colonial sites as well as how "colonial" forms were not spatially restricted to the colonies but were (and are) also part of metropolitan state logics. In other words, the colonial state and the modern state did not develop along two trajectories in isolation from each other. Rather, these coeval formations emerged as jarringly distinct, yet also uncannily similar, through complex *relational* processes, whose contours cannot be captured by understanding the colonial state through the modalities of exception, importation, imposition, or diffusion alone. More broadly, the tendency to attribute the emergence of certain principles, doctrinal propositions, forms of political thinking, and state institutions as autochthonous European inventions that are then applied (or not) to colonial settings misses a crucial part of the story: that such colonial sites were often central to the making of principles, the shaping of doctrine, and the emergence of state institutions and practice. This is not to say that a colonial dimension inhered equally or evenly in *all* aspects of European thought and practice regarding the state; nor is it to say that the vocabulary of importation, imposition, or diffusion is *never* useful in describing phenomena in the colonies. It *is* to say that the analytical and epistemological frames embodied in this lexicon are insufficient to understanding the making of the modern world and must be supplemented by a modality attentive to coproduction.

Departing from a framework that treats metropoles and colonies as distinct, unrelated entities, a growing body of recent scholarship has explored their densely intertwined character, investigating how colonial and imperial concerns shaped developments in the metropoles.[47] As a result, alongside more detailed examinations of various colonized sites, we are better informed on how empire shaped European political thought, on how practices in the colonies found their way back to the metropoles, or on how everyday life and cultural idioms in the metropoles were saturated by colonial–imperial thinking. Under rubrics such as the new imperial history, connected histories, transnational histories, and histories of the global, not only are the connections between metropolitan and colonial sites being evaluated afresh; increasingly, there is more attention to the circulations and webs of connections between different colonies, as also studies organized through different analytical frames that emphasize regional or oceanic connections other than the Atlantic world, such as the Indian Ocean world or the Bay of Bengal littoral.[48]

But investigations of the cultural and ideological traffic between metropole and colony often fall short of exploring the nuances of state formation, particularly in resisting approaches that do not assume its territorial closure. In fact, it is customary to study "the state" as a spatially specified and contained entity. I wish to emphasize that there is, without doubt, much to be learned from site-specific inquiries, and nuanced explorations of several important aspects of state formation and practice entail—indeed, demand—such a perspective. However, limiting analysis to such an approach makes it impossible to follow pathways that can describe state practices whose coordinates are not spatially bound or to analyze state forms that do not adhere to such territorial closure. Migration is one such domain and, as we will see, the regulation of migration is one of the primary mechanisms for the production of the state and of sovereignty as enclosing a fixed territory and a fixed population; for defining membership in political communities; and for consolidating the notion of state borders.[49] In other words, the regulation of migration is a central aspect of what John Torpey calls the "stateness" of the modern state.[50] In tracing a *colonial genealogy of the modern state* this book is specifically concerned with analyzing how the mechanisms and rationales that emerged in the regulation of colonial migrations are embedded in normative understandings of the modern state and how they inform, animate, and, often, are amplified in current migration regimes.[51]

I use here a Foucauldian notion of genealogy, directed toward producing an "effective history": one that illuminates how things emerged and descended, in the form they did—not through the unfolding of a foregone teleology, but through chance occurrences, peculiar configurations, contingent forces.[52] Indeed, as we cover the terrain of the regulation of colonial Indian migration we find that regulations were often made to address contingent exigencies, thus dispensing with any logical consistency that might ground a smooth legal order. As a result, at the heart of a legal regime and a bureaucracy striving for an ideal rule of law that rested on definitional clarity, unambiguous regulations, and systematic implementation lay a thicket of ambiguity and a muddle of laws and rules such that the making of modern state formation took shape, in this instance as in others, in a manner that was arbitrary, ad hoc, but, in these ways, not unique.[53] In much the same way as Antony Anghie demonstrates that the so-called sovereignty *doctrine* emerged through a series of "improvisations" that spoke to particular historical conjunctures, the relations between migration and the state did not flow from pre-given principles; rather, they emerged in haphazard fashion in response to particular historical exigencies.[54] This work analyzes aspects of the particular circumstances, the complex debates, and the novel arguments attendant on state regulation of colonial Indian migration that, importantly, have come to be congealed in routinized technologies and unquestioned ideologies. In doing so, it traces the historically specific and shifting axes along which the state controls migration to illustrate how they are crucially linked to the production of historically specific subjects, to the formation of certain kinds of states, and to the generation of particular kinds of inter-state relations. In this way, what genealogy or effective history offers for serious consideration is the peculiarity and contingency of the present and, hence, the prospect that the present provides *no necessary* template for the future; it thus offers the possibility not of an apologia, but of a thoroughgoing critique of the present. In other words, the colonial genealogy of the modern state I chart here both facilitates an appreciation of the complexities of the historical contingencies that produced the present and provides an avenue for articulating futures that are not merely versions of, or smoothly continuous with, the past.

In concrete terms, this project is a critical assessment of the official archive, or the state's account—one can call it the state's memory, albeit patchy and partial—of Indian migration from about 1834 to 1917.[55] In the space of some seventy years, from 1834 to the turn of the century, indentured Indian migration would be authorized for some twenty countries on four continents. As a result, people who were variously called "Indians," "East Indians," "Asiatics," "coolies," "natives of India," "Hindoos," and so forth, would, over the years, arrive in such disparate locations as present-day Mauritius, Guyana, Jamaica, Trinidad, Réunion, Suriname, Guadeloupe, South Africa, Tanzania, Uganda, Fiji, Australia, and a host of other sites. In each case, state-regulated indentured migration was authorized as an *exception* to the then-prevalent, general principle of free movement. These authorizations, which occurred alongside the *lack* of regulation of "really free, 'free' migration" (a paradoxical formulation, whose confounding complexity I attempt to explicate in chapter 1), would necessitate negotiations between all sorts of actors that included, but were not limited to, the French, British, Dutch, Portuguese, and Japanese empires, and various local princely states in India. Since the regulation of Indian migration, in the period I consider, is concurrent with the expansion of the state apparatus more generally, there is a massive, unwieldy archive whose constituent parts are strewn across the globe and, on average, runs to several thousand pages per year (excluding the loss or planned destruction of, perhaps, a bulk of the documents).

In other words, what we can call the conventional archive on Indian migration is vast; the difficulties of navigating the classificatory rationales of both bureaucratic and archival practice make it impossible to present an adequate picture of this "vastness" or to definitively master its contents. For instance, over the years, in British India alone, the matter of Indian migration would be the charge of a variety of different departments, whose contours were, in turn, in constant flux and indicate the complexities and vagaries of any definitive contemporaneous classificatory schema.[56] In addition to the varied departments of the Government of (British) India explicitly charged with the matter of migration, we can also find materials directly relevant to Indian migration in a host of other correspondence and communication at varied sites, concerning matters covering such seemingly disparate issues as sanitation, prisons, slave manumissions, maritime law, the formation of an international postal service, or the regulation of financial markets and exchange rates. Thus, to use archival parlance, the relevant fonds are no guarantee of an exhaustive compilation of even the partial, mutilated existing sources. But, and this is the more important point, given the contours of the conventional archive I have described, the particular period it traverses, the geographical scale of its concerns, the different authorities (empires and other states) it involved, when combined with other archives, including "secondary" sources and what is called "theoretical" discourse, offer a very rich corpus of materials for writing a colonial genealogy of the modern state that can account for its dispersed spatio–temporal determinations and multiple instantiations.

Trajectory of the Text

Chapter 1, "The Migration of 'Free' Labor: Contracting Freedom," links together the seemingly disjointed events of the abolition of slavery in Mauritius and the Caribbean, the consolidation of colonialism in India, and the ascendance of liberalism in Britain to consider the first widespread incursions of the state in regulating the movement of "free" subjects in relation to Indian indentured migration. The chapter advances two central arguments: one, regarding the broader nineteenth-century transformations in contract law and understandings of freedom; the other, regarding a fundamental change these regulations embodied in terms of state sovereignty. Since indentured labor was transported to replace slave labor, the primary concern animating these early regulations was to ensure that the migration was "free" and distinguishable from the slave trade. The hallmark of state regulation of Indian indentured migration was the appearance of the state-authorized labor contract each emigrant was required to "sign."[57] Indeed, so salient is the contract to this movement that the migrants themselves referred to each other as *girmit* or *girmitya*, terms derived from "agreement," or contract.[58] However, Indian migration did not simply mobilize preexisting understandings of freedom or of the contract. Rather, as this chapter argues, the debates occasioned by Indian migration in the wake of abolition were one crucial site where we witness the rise of "consent" as a definitive element of "freedom," which characterizes nineteenth-century transformations in contract law. Moreover, such state intervention, which required the management and oversight of private contracts as a prerequisite for migration, did not accord with prevailing understandings of the purview of state authority. It thus occurred amid acute contention, debate, and contestation regarding state authority and the very definition of state sovereignty. The chapter details these debates and outlines how state regulation of the movement of "free" labor was ultimately justified by recourse to what the documents call the "necessary ignorance" of the colonized subject. This "ignorance" covered over the paradox that the state intervened in "free" migration precisely in order to ensure that it was "free," even as it effected a fundamental transformation in accepted understandings of the legitimate purview of the state. Indeed, these state interventions, which emerged from the particular nexus of abolition and the material imperative to provide cheap labor to the sugar-producing colonies, constitute a critical moment in redefining the parameters of "free" migration and reformulating the limits of state authority and sovereignty. Traversing these important twin transformations—in the very meanings of a free labor contract and in state sovereignty—is the subject of chapter 1.

Despite the advent of state-authorized contracts, a concern with "freedom" would haunt the system of Indian indenture for the near-century it was in existence. However, alongside the persistent concern with freedom that surrounded indenture, after about 1850 we witness a marked quantitative and qualitative shift in migration regulations: in tandem with the expansion of the state apparatus more generally, there was a proliferation of migration regulations monitoring every aspect of the movement. These changes are the subject of chapter 2, "Disciplinary Power and the Colonial State: The Bureaucracy of Migration Control." Deploying a Foucauldian notion of disciplinary power (as distinct from sovereign power that organized contractual relations, discussed in chapter 1), this chapter charts the formation of a huge bureaucracy of medical workers, health inspectors, police officers, recruitment agents, em/immigration officers, and the like, which emerged from a complex web of transcontinental correspondence and negotiation. My approach and objective here differ from existing scholarship (including my own) that has offered rich accounts of legislative action and other measures that regulated Indian migration and organized the terms of life and labor at the destination colonies. Such work is indispensable to enhancing our understanding of the specificities of Indian migration to certain locales, of the particular contexts that produced one set of legislation and regulation rather than another, of the consequences of different measures, of the varied experiences of migrants, and, more broadly, of a social history of migration. But, in the main, such approaches do not help us grasp the *overarching* contours of a migration *regime*.[59] In this chapter, I examine varied state measures and emerging state logics both to analyze the processes through which a vast and complex migration bureaucracy took shape and to provide a profile of its architecture. The regime was characterized by the dense entanglements of "colonial" and "modern" forms, which were simultaneously instantiations of and catalysts for transformations in the state. This mammoth bureaucracy spawned an array of mechanisms for micro-managing migration whose function was not to ascertain that the emigrants were "free" but to monitor all manner of other variables—from the health and fitness of emigrants to the character of recruitment agents, from the dietary scales on ships to the sleeping arrangements and exercise routines of emigrants. Unlike the contract, which had provoked intense debates on the very definitions of freedom, sovereignty, and the limits of state authority, these regulations did not provoke debates on the legitimate purview of state authority. However, despite the absence of such debates, this chapter analyzes how these changes, in fact, both embodied and reflected the widespread deployment of disciplinary power and constituted new relations between states and subjects.

A frequent misunderstanding in Foucauldian scholarship is that the proliferation of nonsovereign forms of power (such as disciplinary power, governmentality, or biopower) is accompanied with the disappearance, or certainly the retreat, of sovereign power.[60] Given that the contract was a central technology in the management of Indian migration, and it is perhaps the quintessential expression of sovereign power, it is foolhardy to ignore the operations of sovereign power within the regime of Indian migration control. Instead, it will be important to trace the *heterogeneity of power* and attempt to understand the relationship between the different modalities of power, particularly since such relationships might well work differently in the colonial field than in the field Foucault describes. However, despite the European context that Foucault theorizes, aspects of his analysis resonate deeply in non-European locales precisely since the logics of power, the formations of state institutions, the grammars of control, and the modalities that (re)organized the social—in short, what Dipesh Chakrabarty calls "the plural history of power"—were historically, or *empirically*, not segregated or limited to the geographical space of Europe.[61] Indeed, in terms of this empirical unity, the thinking of Jeremy Bentham (central to Foucault's elaboration of disciplinary power) and the Utilitarians informs developments both in Europe and in the colonies, especially India.[62] However, the empirical and analytical unity of metropole and colony should not be understood to mean that the logics of power unfold similarly in all domains; indeed, this is rarely, if ever, the case. Working through the grids of similarity and difference, points of emergence, and patterns of circulation is thus a central aspect of my analysis, as I attend to the *imbrication* of sovereign and disciplinary power to provide a sketch of the overarching migration *regime* that took shape.

Chapters 3 and 4 turn to an analysis of non-indentured migration that, until the early twentieth century, had been largely unregulated by the state. The specific nineteenth-century historical context of the abolition of slavery had precipitated state control of migration along the axis of freedom and within what I have called a *logic of facilitation*. Here, the state-authorized contract was explicitly instituted to *enable* the movement of indentured labor to plantation economies. The twentieth-century context of increasing racial anxieties in white-settler colonies would precipitate state control of non-indentured, or "more free, 'free' migration," along the axis of nationality and within a *logic of constraint*. Thus, even as the global economy and "the economic forces behind migration grew increasingly integrated around the world," migration streams were segregated and divided, based on racial criteria.[63] Premised on racial thinking and contravening a straightforward capitalist rationale, state regulations in the twentieth century would expand to cover a wider array of movement, even as debates over state sovereignty, the purview of state authority, and the relation between states and citizens/subjects would reemerge with renewed strength and take radically new forms. Charting these novel forms of regulation and legitimation, particularly the complex rationales and modes of displacement that shaped the formation of a decisively racialized migration regime, is the focus of chapters 3 and 4.

Chapter 3, "Gendered Nationalism, the Racialized State, and the Making of Migration Law: The Indian 'Marriage Question' in South Africa," traces the emergence of a state-authorized marriage license to detail how an articulation between kinship, nationality, and religion become a central feature of migration control. The particular focus of this chapter is the controversy, between 1911 and 1914, surrounding the legal recognition of polygamous Indian marriage in South Africa. This controversy unfolded in the context of the formation of the Union of South Africa in the immediate aftermath of the South African (or Anglo–Boer) war and became tied to the celebrated *satyagraha* movement spearheaded by Gandhi, who then lived in South Africa. The specific nature of the articulation between the "marriage question" and satyagraha introduced into the calculus a densely gendered dynamic of Indian nationalism with enormous consequences for the terms of the resolution achieved. Aspects of these events in South Africa have received substantial attention in recent scholarship on a global history of migration control.[64] But, in the main, this work has focused on Gandhi's involvement and experiences of racism, neglecting to consider the profoundly gendered dimensions of the events. My analysis here seeks simultaneously to intervene in the historiography of Gandhian satyagraha in South Africa and highlight how race-based migration regulations emerged through a process saturated by the dynamics of a gendered nationalism.

With regard to indentured Indian migration, marriage was seen as an index of good health and sound morality and, for these reasons, largely served as a mechanism facilitating migration. However, with regard to non-indentured migration, marriage was activated as a central institution demarcating the difference between various religiously defined nationalities and functioned as a mechanism constraining mobility. With the newly formed South African state positioned as the representative of a coherent, religiously and racially defined white Christian nationality, migration regulations would increasingly demand that the kinship relations of migrants replicate the Christian nuclear family. Acute and complex questions about the fundamental liberal principles of tolerance and a respect for difference, the separation of church and state, and the demarcation of private and public spheres were resolved by recourse to new definitions of state sovereignty articulated to novel understandings of national security.[65] This linkage enabled vastly expanded notions of security that posited varied kinship relations as a threat to the social fabric of settler societies, thus requiring concerted defenses in the form of migration regulations. By charting the debates attendant on the emergence of the marriage license and its insertion into a regime of migration control, this chapter thus addresses the range of profound transformations that migration wrought upon state formations and inter-state relations that operated at the level of personal kinship affiliations. Currently, it appears that bonds of kinship, particularly monogamous, heterosexual marriage, constitute a chief modality for potential access to mobility; however, the history of the marriage license offered in this chapter demonstrates how the incorporation of marriage and kinship into migration regulations were directed, instead, at restricting movement and represented new axes of state intervention in controlling the mobility of people.[66]

Occurring almost contemporaneously with the controversy over polygamous Indian marriage in South Africa was the controversy—half a world away—in Canada concerning the arrival of increasing numbers of "free" Indian migrants. Unable to prevent this movement within prevailing migration norms and regulations, Canada would become embroiled in a protracted ten-year discussion and debate (from 1906 to 1917) with the Colonial Office in London and the Government of India, directed toward devising a resolution. These debates and the ensuing resolution are the focus of chapter 4, "Race, Nationality, Mobility: A History of the Passport." This chapter undertakes the task of piecing together one history of the passport—the emblematic signifier of modern nationality and (differential) mobility that, *pace* what some herald as diasporic public spheres and hybrid identities, are nowhere in the process of dissolution. Instead, the chapter charts the development of the passport as a document forged to restrict the movement of Indians to Canada to argue, first, that the passport is a crucial mechanism for suturing together discourses of race and nationality; second, that such events as the movement of Indians (and others, especially Asians) to white-settler colonies necessitated the incursion of the state into all mobility, leading to what is now a truism, that (nation-) states must exercise a monopoly over migration practices; and, third, that given a technology such as the passport, modern migration produces nationality. In contrast to work that sees migration as disrupting the contours of national identity, this chapter suggests that migration helps *produce* nationality as a strong territorial attachment. Finally, the analysis here argues that colonial migrations, conceived in racial terms, were fundamental to effecting a transformation in understandings of state sovereignty and to generating a system of nation-states where control over migration is putatively held to be a defining feature of state sovereignty.

The epilogue to this project, "In History: A Colonial Genealogy of the Modern State," consolidates the implications the histories traced in the previous chapters raise for our understandings of the relationships between the colonial state and the modern state. It elaborates on the analytical and epistemological approaches one might adopt to understand these relations and shows that today a colonial dimension is inherent in the modern state globally. Those designated as "migrants" are now the naturalized subjects of a differentiated legal regime within the same territory—the definitive and distinctive feature of colonial state formations. In other words, produced out of the histories I narrate is a modern world that can sustain colonial formations globally—not merely in terms of *inter*-national differentiations, but also in terms of *intra*-national differentiations. In offering a colonial genealogy of the modern state, it is such imbrications and enduring legacies this book seeks to reveal.

CHAPTER 1

The Migration of "Free" Labor

Contracting Freedom

The abolition of slavery has rendered the British colonies the scene of an experiment, whether the staple products of tropical countries can be raised as effectually and as advantageously by the labour of freemen, as by that of slaves. To bring that momentous question to a fair trial, it is requisite that no unnecessary discouragement should be given to the introduction of free labourers into our colonies. So far as it may be inevitable to obstruct such immigration by taking effective securities that the immigrants shall be, in the fullest sense of the term, free agents,—that obstruction may be justified, but no further.

—LORD STANLEY, Secretary of State for the Colonies, to Sir Lionel Smith, Governor of Mauritius, January 22, 1842

The abolition of slavery had not merely "rendered the British [plantation] colonies the scene of an experiment" but, in conjunction with the expansion of colonialism and the widespread deployment of a specifically liberal–capitalist discourse on freedom, had rendered the entire world the scene of that experiment. The hypothesis that the experiment set out to prove was somewhat more ambitious than the limited one Lord Stanley, Secretary of State for the Colonies, intimates in his dispatch: that free labor could be as effective and advantageous to the capitalist as slave labor. On a larger scale, the experiment set out to establish that free labor, or social relations characterized by the commodification of labor power, was the manifest and true destiny of humankind; that, indeed, the "universalization of free labour [was] the *raison d'être* of history."[1] Although the installation of free labor was the initial aim of the experiment, over the course of the nineteenth century it came also to constitute its axiomatic premise. In other words, the precepts of political economy were advanced as the "natural condition of humanity" even as producing and instituting labor relations that adhered to this so-called natural state of freedom became central to the civilizing mission of colonialism itself. In part, then, what was at stake in the success of the experiment of transporting Indian indentured labor to replace slave labor was the very future of liberalism.[2] This future, however, was not one where such historical events simply filled out a shell whose outlines were already drawn; rather, the abolition of slavery and the subsequent system of Indian indentured migration were elements that would crucially shape the future of liberalism and newly emergent understandings of freedom.[3] In particular, as this chapter argues, these historical events would have a lasting impact on the contours of nineteenth-century contract law and on normative definitions of state sovereignty in relation to migration.

Legal historians and theorists have noted that the nineteenth century saw an ascendance of the notion of consent, or "will," to contract law.[4] However, this historical shift has not been related to British slavery abolition, perhaps the most significant legal transformation of labor relations, and hence the terms of contract, in the nineteenth century.[5] James Gordley, for instance, in considering the causes for such a transformation—particularly the eclipse of the concept of "equality in exchange" as well as the philosophical and legal incoherence surrounding notions of "duress," "fraud," and "mistake" to the validity of contracts—catalogues a range of possible reasons: the growth of large-scale markets, the rise of industrialization, and changes in the ideology of nineteenth-century liberalism. He suggests that the centrality of consent to nineteenth-century contract law, championed by the "will theorists," cannot be attributed to any of these factors. By his account, jurists' propensity to accept will theories is best understood as indicative of the decline of an Aristotelian philosophical tradition and an "intellectual crisis within their own discipline."[6] Missing, however, from Gordley's catalogue of possible causes for changes in contract law is the historical event of abolition, as related either to institutional changes in how contract worked or to political, social, and ideological changes in how freedom was conceived.

In another register, historians working on the post-abolition transformations in the legal terms of labor have largely tended to assume that emergent contractual relations of ex-slaves, apprentices, and indentured labor simply duplicated—if often in terms of deviation—extant contractual arrangements that defined free labor.[7] In fact, as Douglas Hay and Paul Craven point out, "much recent work on indentured and other labor in the post-slavery empire remains relatively incurious about the details of the law and its enforcement."[8] In other words, historians often evaluate the contracts made by ex-slaves, apprentices, or indentured labor with reference to a fixed or stable definition of a free labor contract.[9] Thus, where legal theorists have overlooked how abolition might have wrought changes in nineteenth-century contract law, historians working on the changes wrought by abolition have tended to overlook the transformations in what constituted a free labor contract. A central task of this chapter is to read the debates on and definitions of contracts in the regulation of the post-abolition practice of indentured Indian migration alongside the transformations in contract law, to bring into view a different account of the centrality of consent, or will, to nineteenth-century contract law.

A second, closely affiliated, task is to expand discussions on state control of migration, particularly in relation to understandings of state sovereignty. As noted in the introduction, while historians of migration have provided detailed descriptions of the manifold migration controls undertaken by a variety of states, they have not given sufficient consideration to how such controls were profoundly intertwined with the historical development of, and transformations in, the modern state and inter-state relations. Simultaneously, theorists of the state have favored, with few exceptions, a diffusionist or mimetic model of the modern state. As we will see, a politico–historical consideration of the legal debates concerning slavery, abolition, and indenture make such positions increasingly difficult to sustain, and, instead, illuminate colonial genealogies of the modern state.

"This Practice Has No Foundation in Any Existing Law": Colonial (Re)configurations of State Sovereignty

On August 1, 1834, following the British Slavery Abolition Act of 1833, Britain adopted a method for the gradual abolition of slavery in its colonies. The method of gradual abolition bound the recently emancipated slaves to a plantation under a system of apprenticeship for a period of six years, later reduced to four. Thus, the year 1838 marks the formal end of slavery in the British plantation economies. Well before apprenticeship had begun, plantation owners in Mauritius and the Caribbean were already anxious about the feasibility of emancipated slaves providing a continued source of labor on the plantations and, more significantly, were worried about how to expand their enterprises without a concomitant increase in the supply of labor.[10] Hence, in addition to an increase in the illicit slave trade (of both Africans and Indians), the planters in Mauritius had arranged for indentured labor from India.[11] Between 1834 and 1839, that is, the period of apprenticeship, more than 20,000 indentured Indians arrived in Mauritius.[12] In fact, on August 1, 1834, the date Emancipation came into effect, the ship *Sarah* sailed into the harbor at Mauritius carrying thirty-nine indentured Indian migrants.[13]

Mauritius not only was the first colony to initiate the migration of Indian labor but also came to occupy a preeminent status with regard to what was the "great experiment" of the nineteenth century—namely, to ascertain the relative degrees of cost, productivity, and profit of slave versus free labor. It therefore served as a critical site for the success or failure of this experiment, even as practices of indentured migration to Mauritius served as a model to be replicated elsewhere. For instance, in 1836, learning of the system used by the planters in Mauritius, John Gladstone (father of William Gladstone, four-time liberal Prime Minister of Britain) initiated correspondence to have indentured Indian labor shipped to work on his plantations in British Guiana (Guyana). It took Gladstone almost two years to complete the arrangements with the governments of India and British Guiana, the Colonial Office in London, and the shipping agency of Gillanders, Arbuthnot & Co., who was charged with the task of actually recruiting the migrants. The *Hesperus* and *Whitby* eventually sailed from Calcutta (Kolkata) in January 1838 with 437 indentured Indians on board.[14] Only 419 arrived in British Guiana; eighteen had died on the voyage, two of whom had reportedly "[fallen] overboard in a violent gale" and drowned.[15] These Indians came to be known as the "Gladstone Coolies." Slave uprisings on Gladstone's plantations in British Guiana in 1823 had made him notorious with abolitionists in England.[16] His scheme of transporting Indian labor only added to this notoriety.

Coming as it did on the heels of slavery, the migration of Indian indentured labor was, from the start, plagued by opposition, doubt, and anxiety regarding the "freedom" of the migrants. Hence, as early as 1835, when the migration to Mauritius was gaining momentum, the Court of Directors of the East India Company (in charge, at the time, of British administration in India), in an effort to ensure freedom, stipulated that "intending emigrants [were] to appear before a magistrate and satisfy him of their freedom of choice and knowledge of the circumstances of the case."[17] Despite this state oversight of a private contractual arrangement, the anxiety over the freedom of the migrants continued. Thus the government, though recognizing "the inexpediency of throwing impediments in the way of free emigration of the native Indian labourer," had asked the Law Commission in India to give the matter their "serious consideration" and determine "whether any further security can be afforded to these people against ill-treatment than is provided for by the existing orders."[18] The Law Commission felt that extensive legislation and regulation were unwise and unnecessary: "no legislation is advisable, except . . . such as have already been made, to prevent undue advantage being taken of the simplicity and ignorance of [these] persons. . . . If sufficient care be taken to ascertain that every essential point is provided for in the engagements [i.e., the contracts] . . . it does not appear to the Commissioners that there is anything more which the Government of this country can reasonably be expected to do for the protection of [this] class of persons."[19] It was along the lines of such minimalist state involvement that the first piece of legislation regulating the movement of indentured Indian labor was ratified by Act V of 1837. In conjunction with the regulations and the Act in India, in 1835 the government of Mauritius passed two ordinances that, in addition to stipulating the precise terms of labor, prohibited the entry of laborers into Mauritius, "except under the authorization of his Excellency the Governor."[20]

Given that over the course of the nineteenth century the regulation of Indian indentured migration would blossom into a massive, micromanaged, state-controlled enterprise, it is significant that this minimalist state oversight of private contracts constituted a radical departure from prevailing practice. Indeed, in asking for the recommendations of the Law Commission in India, the imperial government had stated: "It is to be observed, that this practice [of regulating migration] has no foundation in any existing law, but was prescribed by [an] order of Government on the first case coming under its notice."[21] Not only was the state aware of the novelty of legal regulation, but the initial regulations, particularly in Mauritius, occurred amid challenges to the state's authority to monitor "free" migration.

For instance, Hollier Griffiths, a planter in Mauritius, writing to G. F. Dick, the Colonial Secretary of Mauritius, found the nature of the Mauritius ordinances to be unprecedented: "[N]o example does exist, to my knowledge, in any civilized country, in modern times at least, of the creation by legislative acts of obstacles to the augmentation of the labour and industry of a community."[22] He objected to the ordinances on the grounds that they would "discourage the introduction into this colony [i.e., Mauritius] of free workmen or labourers," when what was needed was to "promote the establishment of free labourers in this island by means of premiums."[23] But, more importantly, he found it his "duty to question the competency of the authority by which the enactments . . . have been made."[24] Griffiths pointed out that while the sovereign had the authority, in exceptional cases, to prohibit the *departure* of a British subject from British territory, or even to recall him, the royal prerogative "[did] not extend so far as to prohibit the *entrance* into his dominions of any of his subjects."[25] Moreover, wrote Griffiths, "this power vested in his Majesty has been considered so contrary to the liberty of the subject, that very few instances are on record of its ever having been exercised."[26] Griffiths worried that if the government of Mauritius "was legally entitled thus to infringe on the rights of his Majesty's subjects, natives of India, it might equally claim the same authority in respect to the subjects of his Majesty, natives of Great Britain, . . . [who would] thus [be] deprived of a right appertaining to [them] by birth."[27] By passing the 1835 ordinances restricting the entry of immigrants, the local Council of Mauritius had, in effect, superseded even the authority vested in the sovereign.

The relation between mobility, freedom, and the state has received relatively little elaboration in classical liberal theory. Significantly, one of the historical occasions for the limited discussion that exists concerns European migration to the Americas and the accompanying seizure of territories. Such negligible attention might today appear curious; however, it is less curious given the widespread nineteenth-century view that for the state or sovereign authority to hinder the free movement of persons traveling for "peaceful and lawful reasons" was an exceptional, and largely indefensible, exercise of sovereign power.[28] In this regard, the tenets of natural law were incorporated into those of positive law.[29] Thus, for instance, when, in 1792, England exercised exceptional powers to monitor the entry of Jacobins following the French Revolution, it was not without dissent, with contemporaries even denouncing the accompanying Act of Parliament as "equivalent to the suspension of . . . Habeas Corpus."[30] The basis for this exceptional Act was directly related to the *political* threat—almost akin to the threat of an entering army—that the Jacobins were thought to pose to sovereignty. Hollier Griffiths in Mauritius was therefore entirely correct in asserting that there did not exist, at the time, legislative acts on migration that functioned as "obstacles to the augmentation of the labour and industry of a community."

Further, key aspects of liberal theory itself become incoherent if one suspends the premise of free movement. For instance, Locke's notion of "tacit consent" includes an implicit presumption of unconstrained mobility. Locke introduces this important idea in order to circumvent the problem of the lack of *express* consent in his version of a consensual political society. For him, "a child is born a subject of no country or government . . . he is a freeman, at liberty what government he will put himself under, what body politic he will unite himself to," though, simultaneously, an adult who enjoys the benefits and protection of a particular government, "does thereby give his tacit consent" to it.[31] But, continues Locke, "[when an] owner, who has given nothing but such tacit consent to the government, will, by donation, sale or otherwise, quit the said possession, he is at liberty to go and incorporate himself into any other commonwealth, or to agree with others to begin a new one *in vacuis locis*, in any part of the world they can find free and unpossessed [i.e., the New World]."[32]

Contained within Locke's proposition is the ability of the subject to leave the territory of a particular sovereign and enter that of another. Thus, the notion of tacit consent is itself dependent on free mobility, since it is in choosing *not* to leave the territory of a particular sovereign that one tacitly consents to the rules of the state. Within this framework, then, to prohibit the free ingress and egress of people was, in fact, as Griffiths claimed, "contrary to the liberty of the subject." What needed adjudication was how to apportion the relative weight between the principle of the liberty of the subject and the principle of sovereignty that circumscribed these liberties. In the case of Indian indentured migration to Mauritius, the matter was only more complicated since, technically, it concerned what could be construed as the movement of people *within* the domestic bounds of the British empire-state.

Whereas the somewhat cursory attention that political philosophy accorded to the relation between mobility and the state is explicable given the principle of free movement, it is quite inexplicable in a range of recent literature. Thus a common element in such distinct literatures as sociopolitical work on migration, political theories of international migration, or historical and cultural analyses of such markedly different patterns of migration as slavery, indenture, exile, European migration to settler colonies, or postcolonial migrations is the scant attention to how, on what basis, and by what authority the state controls migration. This is not to suggest that scholars have not undertaken the task of describing the conditions of literal travel or that various legislative measures to organize, particularly to restrict, migration have gone unnoticed. Rather, to reiterate a point made in the introduction, it is to suggest that the historically specific and shifting axes along which the state controls migration are crucially linked to the production of historically specific subjects, to the formation of certain kinds of states, and to the generation of particular kinds of inter-state relations. An explicit focus on the legal debates accompanying the formation of some of the technologies that govern migration enables insights into how these technologies emerge from and help consolidate certain understandings of key categories such as free labor or nationality. Such a focus, in other words, works against the belief that grants putative legitimacy to state control over migration and inquires, instead, into its peculiar historicity. Let us return, then, to the debates on Indian indentured migration and consider further how the state would invoke an exceptional authority to monitor the migration and consider also the details of how, why, and with what effects it came to be regulated along the axis of freedom through the technology of the contract.

Where Hollier Griffiths, the planter from Mauritius, was purportedly concerned with how state ordinances infringed on the "liberty of the subject" (we shall see later that his objections rested on less lofty grounds), P. D'Epinay, the Procureur-Général of Mauritius, in an elaborate report responding to Griffiths, raised the issue of the need to protect the Indians. Casting Indian migration as necessarily coerced, D'Epinay opined that "it is not of their own spontaneous movement that those men expatriate themselves . . . it is the contractors of the Mauritius, speculators, who go in quest of them, and drag them from their home, under the enticement of greater wages than they can obtain where they are."[33] Deploying the paradoxical rationale of the rule of colonial difference, D'Epinay *both* affirmed that British Indian subjects had the same rights as "those who reside in any possession, territory, or dependency of Great Britain" *and* questioned whether "the term British subject, and the privileges attached to it, are not according to places and circumstances, susceptible of important division and modification."[34] While D'Epinay showed a concern with the welfare and protection of the Indians, he justified the Mauritius ordinances on the grounds of the special circumstances that obtained with the end of slavery and the particular colonial situation: "It is a distinction common to every metropolis, that their colonies are governed . . . by special laws, because the elements of society are not the same therein as in Europe (especially when slavery existed); [hence] the same system of legislature is not applicable."[35] But despite the fact that slavery had now been abolished, the ordinances were, according to D'Epinay, "a measure of foresight and of internal police, the object of which was not to permit the indiscriminate introduction of persons [into the colony], without morality or means . . . [since] it was felt that such a promiscuous introduction would be more calculated to sow tumult and disorder than to [produce an] increase in [the] industry of the country."[36]

The careful selection of immigrants into Mauritius, D'Epinay explained, was crucial to maintaining "public tranquility"; it was especially "prudent" to take "precautionary measures," such as those embodied in the ordinances, to prevent the arrival of "the refuse of the Indian bazars [*sic*] and with such the germs of disorder."[37] This was particularly vital as the period of apprenticeship approached its end, for it was important that the Indian population serve as an example of industriousness and disciplined labor to the ex-slave apprentices:

[W]ise and prudent precautionary measures [should] be taken . . . when this new population is put into immediate contact with the new apprentices just emerging from slavery, still susceptible of every impression; and to whom it is of importance, at the first step towards civilization, to give [an] idea and examples of order, labour, discipline. This end would be frustrated, if permission were given to associate them with all the vagabonds and all the idlers with which India swarms. . . . Who can say what influence this medley of individuals, with their manners, their usages, and their vices will have on our indigenous population, especially when it shall become wholly free? . . . It is the part of a wise Government to give to it serious attention; it is, therefore, necessary to proceed with caution in the new order of things.[38]

D'Epinay's comments are remarkable for what we can now understand as their prescience. In their invocation of such elements as "internal police," the notion of "public tranquility," the protection of public order, or the idea of the state managing and orchestrating the discipline and development of its subjects in particular directions, they already embody new understandings of danger, protection, and security that would be widely mobilized, decades later, to reconfigure the terms and limits of state sovereignty, particularly in relation to migration control. However, D'Epinay offered these understandings of danger, protection, and security not as a *general* rationale for the regulation of migration, but as a rationale for the *exceptional* ordinances in Mauritius, warranted by the *exceptional* occasion of abolition combined with the *exceptional* case of the colonies, which necessitated special legislative measures. In other words, state authority for intervening in free migration was not deemed to be a general feature of sovereignty, but was recognized as an exceptional measure.

Though the issue of such exceptional measures is addressed most directly in the exchange between Griffiths and D'Epinay in Mauritius, it also framed correspondence emanating from the Court of Directors of the East India Company in London and the Law Commissioners in India that, while instituting regulatory measures, simultaneously recognized that they had "no foundation in any existing law." Moreover, the remarks in the correspondence surrounding the acceptance of these earliest regulations at each site—that is, in India, Britain, and Mauritius—readily admitted that they violated the fundamental "liberty of the subject." The regulations were justified on the basis of the "special circumstances" that obtained with abolition and the status of colonized subjects and, paradoxically, on the basis of protecting the very liberty of the subject they violated. Thus, at one end, in India, state supervision of contracts was deemed necessary to protect the Indians from abuse; at the other end, in Mauritius, restriction on the entry of immigrants was deemed necessary to protect the apprentices. However, what emerged as an exceptional discourse of protection would later become the general discourse underwriting state authority in monitoring migration and reconfiguring the definition of state sovereignty. Indeed, if D'Epinay's description of new kinds of threats and thus new measures of protection, which were concerned with the management of the population, resonate with us today, it is because the understandings of danger and protection precipitated by these exceptional events would later be incorporated into the general framework of state control of migration. In short, state control of migration and attendant definitions of state sovereignty have a crucial colonial genealogy.

If the special circumstances of a post-abolition colonial order made it possible to rework the terms and limits of sovereign authority with regard to migration, this in itself did not prescribe the precise form the newly justified interventions to protect Indians should take. The form of the intervention related less to understandings of state sovereignty than to the terms and definitions of the free labor contract. In point of fact, due to the terms of the prevailing indenture contracts, it turned out that both Griffiths's objections to the ordinances passed by the Council in Mauritius and D'Epinay's defense of them would be moot: Lord Glenelg, then Secretary of State for the Colonies, disallowed the ordinances, agreeing with the objections of critics of indenture that the contractual arrangements they contained subjected the laborers "to restraints and penalties of so extremely onerous a nature as nearly to revive, under a new name, the former servile condition of the great body of the people."[39] Glenelg admonished Sir William Nicolay, Governor of Mauritius, expressing his "deep concern" "that such enactments could obtain either the concurrence of the members of the Council or your own sanction." "The freedom contemplated by Parliament, and to procure which such costly sacrifices have been made [slave owners were paid compensation to the tune of £20 million]," wrote Glenelg, "must be ill understood at Mauritius, if it be supposed to be compatible with such regulations as these, or with any others which may be conceived in a similar spirit."[40] Glenelg was thus less concerned with how the regulations affected the principle of sovereignty than he was with the content of the regulations and the free labor contracts.

Regardless of his stance in this case, a year later Lord Glenelg made a special exception to a British Guiana labor ordinance, which restricted contracts to three years, by approving John Gladstone's venture of transporting Indian indentured labor to British Guiana on five-year contracts.[41] It was an ill-fated exception, for even as the first batch of Indian laborers set sail from Calcutta in 1838, in England, the British Emancipator decried Glenelg's exception to the British Guiana ordinance.[42] This led to impassioned debates in the House of Lords and resulted in requiring Glenelg to call for an inquiry into the condition of the "Gladstone Coolies" in British Guiana.

Just six months later, on July 10, 1838, a public meeting held in Calcutta concluded that "the system [of indenture] was radically bad, and contained in itself the elements of a new species of slavery" and needed extensive scrutiny.[43] Alongside this determination, European merchants in Calcutta "connected with the trade of the Mauritius" sent a representation to the government stating that "nothing [could] be more unjust . . . than the analogy said to subsist between the now abolished slave trade and the free labour market established between those we represent [i.e., the planters in Mauritius] and the natives of this country." Moreover, they claimed that even "if there be truth in the allegations . . . no excuse is thereby formed for the summary suspension of the free labour trade."[44] As a result of this furor, the government ordered an inquiry in India into the "alleged abuses of the existing system," put an immediate stop to migration to the West Indies, and, in November 1838, suspended all migration to Mauritius as well.[45] In May 1839, this suspension took legislative form in Act XIV of 1839 that prohibited all migration under contract.

Thus ended the first attempts to conduct free migration under the aegis of the state.

Contract and Consent: The Post-Abolition Discourse on Freedom

Like Glenelg's reservations, the public outcry generated by these early migration regulations was largely unconcerned with the reconfigurations of state sovereignty they represented. Instead, the chief object of concern was the very matter that had occasioned the regulations in the first place: namely, the gloomy specter of the slave trade and the "freedom" of the migrants. Thus, as apprenticeship came to a close and presented, for the first time, an opportunity for the ex-slaves to engage in a contractual relation, the definition of the contract itself would become a point of contestation through a debate over freedom. The contractual arrangement that governed indentured Indian labor was one site for this contestation.

A six-member committee was appointed in Calcutta to investigate the situation of Indian indentured labor. Theodore Dickens, who had sugar plantations in India, chaired the committee. Its other members were Reverend James Charles, a reformist active in India; Russomoy Dutt, the first Indian to be appointed a Judge of the Small Causes Court; William Dowson, a merchant directly engaged in shipping indentured labor to Mauritius; James Grant, a member of the Bengal Civil Service; and Major Edward Archer.[46] Between August 1838 and January 1839, the committee examined some thirty to forty witnesses, gathered written documents and correspondence, and eventually submitted a report on October 14, 1840.[47] However, when it was submitted, the report was signed by only three members of the committee: Dickens, Charles, and Dutt. Dowson and Grant submitted separate minutes not in consonance with what was called the "majority report." Major Archer, meanwhile, departed from India midway through the work of the committee. His comments appear to have been buried in the morass of official correspondence and did not inform discussions in India.[48]

The authorities in India, England, and the various plantation colonies had awaited the report for more than two years, during which time the migration of Indian labor had been in abeyance. While the split in opinion of members of the committee undermined the authority of the report, together the majority report (also called the Dickens Committee Report) and the dissenting minutes from Dowson and Grant touch on every issue of relevance to the migration of Indians—from the belief in bourgeois political economy to awareness of its dark underside, from the representation of India as overpopulated and "swarming with idlers" to the glories migration offered Indians, from the representation of the ex-slaves as "indolent" to their temerity in demanding high wages for their "free" labor, from the hardships imposed on planters denied access to Indian labor to how this distorted the "free labor market," from the attenuation of the "moral influence" of Britain if it reopened indenture to how the Empire itself would be ruined if it did not, from how the evidence of Europeans was inherently superior to "native evidence" to how the "natives" themselves were the best judge of the situation. Despite this range of issues that jostled around the subject of the migration of Indians, at the heart of both the majority report and the minutes by Grant and Dowson was the fundamental question of freedom and the ability of the contract to serve as its guarantee.

The majority report, or the Dickens Committee Report, signed by Dickens, Charles, and Dutt was an unmitigated condemnation of every aspect of indentured migration. Regarding the conditions in India, they found it "distinctly proved beyond dispute, that the Coolies and other natives exported to the Mauritius and elsewhere were (generally speaking) induced . . . by misrepresentation and deceit"; that, in most cases, "the parties were really incapable of understanding the nature of the contracts they were said to have entered into, even when an opportunity had been afforded apparently sufficient for the purpose"; that "kidnappings prevailed to a considerable extent"; that "an impression was successfully created and maintained . . . that they would be liable to penal consequences if they expressed dissatisfaction at being sent on board ships"; that a similar threat of penal consequences "seems to have induced them . . . to suppress the mention of their grievances [on their return]"; and, finally, that "the legislative enactments and regulations of police made and passed for the prevention of abuses [in India] . . . were of very little effect."[49] In legal terms, one might say that the majority report condemned a system in which fraud and duress were pervasive, thus nullifying the validity of the contracts.

The report attributed much of the misrepresentation and coercive methods to the "native crimps" and recruiters "employed by European and Anglo-Indian undertakers and shippers, who [in turn] were mostly cognizant of these frauds."[50] But even as a system of fraud and deceit was said to saturate indentured migration at every level, the report concluded that "no system . . . would ever suffice completely to counteract the tricks and falsehoods that would be resorted to in India by the Duffadars, Arkottics [i.e., native recruiters] and other such persons engaged in similar avocations."[51] A combination of the trickery and treachery of the native recruiters and the innocence and ignorance of the potential migrants thus made it impossible for the contractual arrangement to function. The source of the problem, in other words, lay in the natives, not in the contractual model of freedom. The report, therefore, recommended that if indentured migration were to continue it could only be under an exceedingly elaborate and (deliberately) financially prohibitive system of "impartial" state control.

While the Dickens Committee Report characterized indentured migration as a system that, in general, did not facilitate consent unsullied by fraud, William Dowson's minute placed the entire blame for misrepresentation on the native recruiters who were responsible for initial recruitment. He conceded that it could well be that "the lower agents . . . told deceptitious stories to the recruits, or in various ways endeavoured to cheat them" but this did not warrant the report's "sweeping assertion" that "whatever amount of deception was practiced, in the first instance . . . was contin-ued to be practiced on them throughout."[52] Despite the fact that "in some points they [i.e., the migrants] are defrauded by those native agents from whom at first they derive the information [regarding migration]," Dowson claimed that "their going is still their voluntary act." In his view, the re-port's suggestion that the migration be prohibited was "to proclaim a prin-ciple in the face of all sound political economy, and, what is worse, [it is] a principle destructive of the liberty of the subject . . . a species of tyranny not for a moment to be endured, as it would destroy the political free agency of the subject."[53] For Dowson, moreover, the "degree of misrepresentation" practiced in indentured recruitment was "common to all projects of the kind, to enlistment [in the military], to the manning of vessels, to the in-ducing [of] emigrants to go from the United Kingdom to America or the colonies," and thus "it is not fair to deal with it . . . as if it were peculiar to and exclusively characteristic of the Coolie labor system."[54] Unlike the Dickens Committee Report, which characterized the natives as incapable of consent, Dowson's minute represented the natives as capable of consent and "political free agency."

Hence, according to the Dickens Committee Report, the system of indenture impinged on the liberty of the subject and could be rectified only with extensive, and unfeasible, state intervention, while, according to Dowson's minute, the infringements prevalent in indenture were no dif-ferent from a range of other activities, such that to regulate the migration was to endorse "a principle destructive of the liberty of the subject." At issue, once again, if in different form, was the dilemma we have encoun-tered before with regard to the correspondence, five years earlier, between Griffiths and D'Epinay, namely, how to weigh the principle of sovereignty against the principle of the liberty of the subject. The demarcation was crucially dependent on the particular historical conjuncture that obtained, whose significant elements were the abolition of slavery, the (expanding) colonization of India, and the novelty and unprecedented nature of state regulation of "free" migration. What we are attempting to chart here is the specific colonial/modern trajectory state intervention would take. Given the status of documents such as the Dickens Committee Report and Dow-son's minute, it is useful to further interrogate their contents in some de-tail, for they are significant sites for the consolidation of colonial discourse and for tracing the colonial origins of regulating "free" migration.

The Dickens Committee Report had, in part, arrived at its conclusions by admitting as evidence the testimonies of returned migrants, stating that "after all due allowance [is] made for the habit of exaggeration prevalent among Bengalees [sic] and Hindoostanees [sic]," "it must be admitted [that], on such questions, these persons, however ignorant, are really the best judges."[55] In contrast, Dowson's minute, which ceded political free agency to the natives and drew a direct parallel between the situation of colonized subjects in India and subjects in the metropole, would have no truck with thus equating native and European testimony:

> If by the general term "evidence," which my colleagues use as if it embraced the whole testimony, they include and put faith in such a deposition as that of Abdoolah Khan [a migrant sent back for "misconduct" that constituted his refusing lower wages than he had been promised] . . . they are weighing the value and credibility of evidence in a very different manner from that which is laid down as the proper method and principle by all writers on the subject. . . . [M]y colleagues well know no value can be put . . . [on the section of the report that] rests on native evidence . . . in opposition to European.[56]

Though the Dickens Committee Report and Dowson's minute had entirely different views on the future of indentured migration, they reveal the same lacunae and regularities in the morphology of colonial thinking. Where the report admitted the testimony of the natives, albeit making allowances for their "ignorance" and "habit of exaggeration," to conclude that the system of indenture did not permit consent, *even* in those cases "when an opportunity had been afforded apparently sufficient for the purpose," Dowson refused to countenance native testimony as constituting "evidence," while advocating for the "liberty of the [colonial] subject." At work, then, in these documents is what Homi Bhabha calls the "ambivalence of colonial discourse"—a discourse that institutes the stereotype, supposedly characterized by fixity, even as the motor that drives this formulation is ambivalence.[57] Thus the stereotype is always in excess of what can be empirically proven or observed and also always short of what might be represented. Vacillating between the poles of duplicity and deceit, on the one hand, and innocence and ignorance, on the other, is the ambivalent figure of the colonial stereotype.

If these are the lacunae of colonial discourse, the regularities consist in the larger discursive field that the Dickens Committee Report and Dowson's minute inhabit, a discursive field that activated a specifically colonial understanding of freedom. The duality between freedom and unfreedom/slavery is not merely the axis on which the writers of either document selected to map the events and practices they confronted. The very form and shape of the regulation of Indian migration following in the wake of slave labor dictated that this was the only available axis. For the report, since the conditions of indenture did not facilitate consent they were patently similar to those of slavery; this was cause to discontinue the system of Indian migration. For Dowson, reports of fraud and misrepresentation were insufficient reasons for a "summary suspension of the free labour trade." But, for both, it was what the report called "the character of the natives" that stymied the functioning of freedom and the contract.[58] Moreover, the centrality of consent as a preeminent defining feature of the contract was not itself at issue, just as the contract as a technology for instituting freedom did not come under scrutiny.

The preponderance of the free/unfree duality and the centrality of the contract is not unique to these documents. It characterizes the entire official archive on Indian indentured migration and has framed most historiographies of the phenomenon.[59] For instance, Hugh Tinker, in his pioneering, classic study of Indian indentured migration, writes that he initially approached the material hoping that by examining a variety of positions on the matter "'objectivity' [would be] preserved by taking a median stance between the extremes," but with the "accumulation of evidence . . . [reached] the conclusion that indenture and other forms of servitude did, indeed, replicate the actual conditions of slavery."[60] More than twenty years later, David Northrup found that his view "shifted from near Tinker's final outlook back toward a median position, which sees indentured labor overall as having more in common with the experience of 'free' migrants of the same era than with the victims of the slave trade."[61] Tinker's position, therefore, is allied with that of the Dickens Committee Report, while Northrup's echoes Dowson's position.

Despite the fact that more than a century separates the documents in the conventional archive and these historiographies, Tinker and Northrup accept, rather than interrogate, the terms set forth by the contemporary proponents and opponents of indentured migration, unaware of the historicity of a discourse on freedom wherein slavery (signifying violation, coercion, unfreedom) appears at one end of the spectrum and free labor (signifying volition, consent, freedom) at the other. The mechanism that produces these dualities of volition/violation, consent/coercion, freedom/unfreedom, namely the contract, escapes examination. Left unquestioned is the very practice of thus arranging and quantifying freedom within a teleological narrative such that the category of free labor emerges as an ontological fact that exists outside relations of power and, indeed, outside history itself.[62] Thus, as with the contemporary accounts, which were concerned with ascertaining the freedom or unfreedom of the migrants, historians too have been preoccupied with the impossible task of ascertaining the degree of the migrants' consent.[63] Since the terms of the contracts were so appalling, Tinker concludes that the migrants could not have consented to them; on the other hand, since the terms were somewhat similar to a range of other contracts, Northrup accepts the consent of the migrants. Neither, however, questions what relation the contract bears to freedom. In inquiring into how consent became so central to notions of freedom and to the free labor contract, in particular, I am thus seeking here to also historicize our own investments in consent.

The matter of consent was also at the heart of the concerns of committees established to investigate the situation of Indians in Mauritius who had arrived there prior to the 1838/1839 suspension of migration. The primary charge of these committees was to determine if the Indians had consented to the terms of the "engagements" they had entered into and ascertain if the terms of such contracts were upheld. The committees designed a survey consisting of a list of questions to guide their inquiries of the Indian laborers at each estate. In addition to questions regarding the quantity of food provided and whether the laborers were paid regularly, the survey included questions such as "Whether they are satisfied with the performance of the engagement on the part of their employer, and do they clearly understand its conditions?" or "Whether they were originally engaged in India on account of their present employer, or transferred to him by other parties?" If so, was such a transfer made "with their free consent?"[64] "When required to perform extra work are they paid for it; and whether they do so with their own free will?"[65] The answers to these complex, metaphysical questions of consent and will, in keeping with the survey format, were perfunctory and recorded in the following ways: "Yes; they comprehend their engagement and its conditions," or merely as "Satisfied," or "Yes," or "They are obliged to work till eight A.M. on the Sunday, without receiving any payment for the same," or "Yes; when they work on Sunday they are paid for it; they do it voluntarily."[66]

The report submitted by the first five-member committee appointed in Mauritius, while including in appendix form the survey responses from the laborers, further telescoped the already truncated survey format into an "abstract," consisting of a three-page table that summarized findings from almost fifteen hundred laborers on thirty-one plantations.[67] As in India, in Mauritius too the inquiries resulted in dissension among the members of the committee. One member, Charles Anderson, refused to sign the abstract summarizing the committee's findings, which, in his view, was "calculated to mislead any person who might have recourse to the abstract for the purpose of forming a judgment on the question at issue."[68] Anderson's evaluation of the system diverged from that of other members: on the one hand, he was convinced that "the Indian labourers [were] employed . . . [and] generally fed, clothed, and paid with but little deviation from their agreements"; but, on the other hand, he was also convinced that the laborers were, at each stage, "treated with great and unjust severity":

> Many of them have actually been kidnapped from their own country, where they have been induced to leave under circumstances of gross fraud; . . . They reach this colony . . . [where] they are compelled, by the engagements to which their own ignorance and the avarice of others has bound them, to toil during five years *for a recompense bearing no proportion to the work to which they are subjected*, when compared with the common estimation of the value of labour in this colony, or to the sum which they would earn if they had the free disposal of their time.
>
> The harsh treatment which has been adopted . . . cannot in any shape be justified . . . [and] must show the necessity of applying a sure and speedy remedy, if free labour is to be placed on that footing which sound policy, justice, and common humanity would dictate.[69]

While, at first glance, Anderson's criticism of the system, particularly its emphasis on fraud, seems similar to that of the Dickens Committee Report in India, there are crucial differences. Anderson drew direct attention to the unfair terms of the contract, though he conceded that the terms of the contract were met. Further, while concerned with degrees of fraud and misrepresentation involved in recruiting migrants, he did not attribute this to the "character of the natives," and, in his view, even if the migrants consented to the contracts, their terms were, nonetheless, unfair. The Dickens Committee Report, on the other hand, was centrally concerned with how fraud and duress vitiated consent and had framed its objections to indenture as frustrating this central aspect of the contract. This distinction is an important one, for these differences embody and evince a critical, nineteenth-century transformation where an obsessive concern with consent or will would become the most significant feature of a contract.[70] Although earlier definitions of a contract certainly included a concept of consent or will, they also gave weight to several other important elements such as equality in exchange and the absence of fraud, duress, or mistake. If a contract did not fulfill each of these criteria, law could deem it invalid. Prior to the nineteenth century, simple consent to a patently unfair contract did not guarantee its validity. In other words, consent in and of itself did not make a contract binding.

Over the course of the nineteenth century, there were two profound changes to the understandings of a valid contract. First, a contract was defined as binding and enforceable in terms of will or consent; this was a circular argument: a contract was binding because one willed it and if one willed a contract it became binding.[71] Second, the principle of equality in exchange was discarded as paternalistic and unnecessarily interventionist.[72] Within this model, the "fairness of a contract could no longer be analysed . . . The will of the parties had been separated from, and indeed opposed to, concern about the fairness of their contract."[73] The will, moreover, "had to be declared outwardly since . . . willing is an 'invisible event.'"[74]

If there is some consensus on the specific elements that characterize nineteenth-century changes to understandings of a contract, there is much debate over when, precisely, these changes occurred or how we might explain them. Some legal historians have noted that these transformations accorded well with nineteenth-century liberalism, particularly with notions of legal individualism and economic liberalism.[75] These, in turn, have been linked to the growth of markets and the rise of industrialization, which facilitated the formation of contract doctrine more sympathetic to the commercial classes.[76] Others, such as Gordley, argue that neither social and economic transformations, such as industrialization, nor appeals to changing ideas and philosophies are adequate explanations. He suggests that the doctrinal changes in contract law reflect the decline of an Aristotelian philosophical tradition, which produced an incoherent legal framework, wherein consent reigned supreme and rendered puzzling any limits on consent, such as those embodied in notions of equality in exchange.[77]

In these varied accounts, conspicuous by its absence as a crucial *site* (if not a possible *cause* or *explanation*) for this reconfiguration is abolition, the core element that structured the debates over the contracts made by indentured Indians.[78] This is a result of the narrow national framework that organizes much legal history and the propensity to understand colonial and metropolitan legal regimes and state formations as either following distinct trajectories or as confined within the imposition/importation model conceived as a one-way traffic from the metropole to the colony.[79] If we alter the analytical framework, a colonial genealogy of the modern state comes into view to challenge the idea that the metropolitan state is unaffected by events, including legal events, that occur in the colonies and to question the dispersal model of state formation. Pursuing this vein of analysis, I am struck by how the correspondence I have been detailing was engaged precisely, if perhaps unknowingly, in rethinking and remaking the terms of valid contracts.[80] These debates, moreover, occurred contemporaneously with, if not slightly prior to, the debates of jurists in France, England, Germany, and the United States.

Debates over slavery provided the occasion for some of the most extensive discussions on freedom in the nineteenth century. Given this, is it possible that the aftermath of British abolition fundamentally shaped the changes in that most liberal of institutions, the free labor contract? Might it be that the historical phenomenon of British abolition provides the best instantiation, if not explanation, for the near-exclusive centrality of consent and the accompanying decline of a concern for equality in exchange? Could it be that the ascendance of consent as the defining element of contract law took shape not chiefly in doctrinal pronouncements or in recorded case law but in bureaucratic and political wrangling and in enforcement rather than statute, what Hay and Craven call "low law"?[81] Last, could it be that abolition and its aftermath, including the specificities of Indian indentured migration, are an important part of the archive of the expansion of contract law to extend beyond property, traditionally conceived, to include domains such as the labor contract?[82] In order to resolve these questions, we turn to the discourse of bourgeois political economy that framed the discussion over the commodity—and property—of labor power.

"Incontestable Principles":
Political Economy and the Legal Subject of Capital

In his classic study *The Political Theory of Possessive Individualism*, C. B. Macpherson argues that English liberal thought from the seventeenth to the nineteenth centuries shares an assumption of what he terms "possessive individualism." Within this thought, the individual is conceived as "essentially the proprietor of his own person and capacities" and as one who, although he "cannot alienate the whole of his property in his person, may alienate his capacity to labour."[83] Possessive individualism is the basis for what Macpherson calls a "possessive market society" wherein all human relations take the form of market relations of exchange.[84] Such a society requires a strong sovereign who, to use Hobbes's words, "appoint[s] in what manner, all kinds of contract between Subjects, (as buying, selling, exchanging, borrowing, lending, letting, and taking to hire,) are to bee made; and by what words, and signes they shall be understood for valid."[85] In moments of transition, for instance from a customary society to a possessive market society, the need for a strong sovereign, writes Macpherson, is particularly crucial.[86]

Macpherson's possessive market society is, as he acknowledges, similar to Marx's capitalist society, one of whose singular distinguishing features is the widespread appearance of labor-power as a commodity on the market. As Marx writes, in order for labor-power to appear as a commodity on the market,

> its possessor, the individual whose labour-power it is . . . must have it at his disposal, must be the untrammeled owner of his capacity for labour, *i.e.*, of his person. He and the owner of money meet in the market, and deal with each other as on the basis of equal rights, with this difference alone, that one is the buyer the other the seller; both therefore equal in the eyes of the law. The continuance of this relation demands that the owner of the labour-power should sell it only for a definite period, for if he were to sell it rump and stump, once for all, he would be selling him-

self, converting himself from a free man into a slave, from an owner of a commodity into a commodity.[87]

Distinguishing a slave relation from a capitalist relation, therefore, are the continued and endless iterations of the contract that prevent the free person, the *owner* of labor-power as a commodity, from being converted into a commodity herself.

The resemblance between the analysis Macpherson arrives at, through a reading of English liberal thought, and the one Marx reaches, through a consideration of political economy, need not be belabored. I have referenced both here since Macpherson's focus allows us to locate possessive individualism as the central strand of liberal thought, while not requiring "any particular theory of origin or development of [a possessive market] society."[88] Marx, in turn, is indispensable precisely since his analysis is directed at a critique of the axiomatic assumption of political economy, which "regards the labour-market as [merely] a branch of the general market for commodities," as well as of the ideological matrix it produces and within which it flourishes.[89] Thus Marx writes, with substantial irony, that the sphere of the market, "within whose boundaries the sale and purchase of labour-power goes on, is in fact a very Eden of the innate rights of man":

> There alone rule Freedom, Equality, Property and Bentham. Freedom, because both buyer and seller of a commodity, say of labour-power, are constrained only by their own free will. They contract as free agents, and the agreement they come to, is but the form in which they will give legal expression to their common will. Equality, because each enters into a relation with the other, as with a simple owner of commodities, and they exchange equivalent for equivalent. Property, because each disposes of what is his own. And Bentham because each looks only to himself.[90]

Marx is here describing in 1867, as abolition occurred across the globe, precisely the social formations and new contractual relations, crystallizing around the notion of free will and consent, that had emerged some thirty years earlier within the regulation of indentured Indian migration. Within the documents I am examining, assumptions of Macpherson's possessive individual and the discourse of political economy Marx describes were palpably and explicitly alive and, despite some dissension coming from the

Dickens Committee Report and the communications of Charles Anderson, such understandings were "incontestable."

The Dickens Committee Report presented three primary objections that made allowances for the material conditions in India and Mauritius to question the "abstract principles, as detailed by some political economists, [that] it is always a benefit to a country, where labour is dear to import the commodity from one where labour is cheap."[91] First, it challenged the fairness of having Indians sign contracts and thus bind themselves to a certain wage, *prior* to their departure, when "free labourers, negro or mulatto," in Mauritius, commanded a higher wage.[92] The Report thus suggested that contracts made within India be discontinued. Second, the Committee saw the contracts as not only unfair to the Indians, but the migration itself as explicitly directed at driving down the wages of "free" labor in the colonies. The question for it therefore was "whether an African free labourer, who was taken to Mauritius as a slave, ought not to be protected from any kind of competition than that which arises from really free emigration of other labourers."[93] Against the view of political economy and the free market where human labor was but another commodity to be exchanged on the market, the Report offered the example of the opposition that would follow in the wake of reducing the wages of English agricultural labor through the introduction of Irish laborers on fixed wages.[94] Hence, it questioned the very basis of political economy and the Ricardian principle of comparative advantage that posits that such a movement of labor would benefit *all* parties concerned.

Finally, in the view of the Dickens Committee Report, reopening indentured migration would make arguing for the abolition of slavery in non-British colonies ever more difficult:

> It seems to us that the permission to renew this traffic would weaken the moral influence of the British Government throughout the world, and deaden or utterly destroy the effect of all further remonstrances and negotiations respecting the slave trade . . . for our government would have no reason to urge for remonstrance or interference that could not be answered by a reference to its own example, and on general and abstract principles, . . . that it is the right of all men to trade in free labour, and especially of him whose only property is his capacity to labour, to sell that commodity at the best profit.[95]

The resolution to the difficulties posed by dissension among partici-pants of these first inquiries will lead us to the words I quoted to begin this chapter: the important letter from Lord Stanley, Secretary of State for the Colonies, ordering the resumption of indenture in 1842.[96] The dissent from Charles Anderson, sole champion of the oppressed in Mauritius, turned out not to pose much of a problem. He would take a yearlong leave of ab-sence from his position as Stipendiary Magistrate and begin service in the employ of the planters in Mauritius, lobbying for the continuation of In-dian indentured migration in London.[97] Since, in his initial view, the situ-ation of Indian laborers "required immediate remedy, and merited marked reprobation," Anderson's appointment as the agent for the planters was held, in the Governor's estimation, to "strongly indi[cate] that [the plant-ers'] project is founded on good faith, and an earnest regard for the welfare of the Indians."[98] Moreover, as the preliminary report from the committee in Mauritius had offered an assessment of the "general good treatment of the Indian labourers," it was only the Dickens Committee Report, in India, that constituted an obstacle to reopening indenture.[99]

This Report became the focal point for dissenting minutes by two com-mittee members (Grant and Dowson), for extended debates in Parliament, and for the wrath of planters in Mauritius who, in the meantime, had orga-nized themselves disingenuously, if fittingly enough, as the Mauritius Free Labour Association. The cumulative force of the views of James Grant, William Dowson, the Mauritius Free Labour Association, the supportive members in Parliament, and the changed position of Charles Anderson, combined with the crucial ingredient of the material necessity for the availability of cheap and stable labor in Mauritius, would demolish every reservation of the Dickens Committee Report and, essentially, dictate the resumption of Indian indentured migration.

To the Report's recommendation that the requirement for contracts made in India be abolished and only "really free" emigration permitted, the Mauritius Free Labour Association replied:

Such we consider to be incontestable principles[:] . . . The labour of the Coolie is his merchandise or commodity, and he has a right therefore to dispose of it at the best market, and at the most advantageous price he can obtain. If, therefore, he brings his labour to this market, at his own expense, perils, and risks, he is entitled to obtain to his profit the highest price it affords; but if he requires the aid of another person's capital for the purpose of bringing his labour to this market, he is no more entitled to the whole of this price, than the consignor of any other merchandise is entitled to have the freight, insurance and charges, at the expense of the consignee.[100]

To the Report's charge that the migration of Indian labor on contract would serve to drive down the wages of African ex-slaves, paralleling the hypothetical situation of the wages of English agricultural laborers being reduced through the introduction of Irish contract workers, Dowson responded:

[Such a case] is too superficial to stand a moment's investigation. . . . Why the very fact of their being a people of England shows the inapplicability of the case to the Mauritius, where there are no aborigines, and no peasantry at all having any right to claim a monopoly of the market. . . . In the supposed case, undoubtedly the legislature would interfere to save the English peasantry from the destruction . . . but the labourers who inhabit the Mauritius are in no such position, but in that of men having no patriotic right whatever as children of the soil and whose object and practice it is to keep up the price of labour at a monopoly rate altogether ruinous to the commerce of the island: between the cases there is no analogy.[101]

I leave to the reader the question of whether there existed an analogy between England and Mauritius, regarding the "monopoly" labor exercised in each. Given the fact that indentured migration was resumed, we know that the legislature did not step in to "save" African labor from "destruction," as, Dowson believed, it "undoubtedly" would have done in the case of the English peasantry.

And, finally, to the concern the Report expressed regarding the difficulties reopening indenture posed to the argument for abolition in non-British colonies, came a lofty dismissal from James Grant (later Sir James Grant), member of the Bengal Civil Service:

The argument [of the Report is] . . . that if we permit a free circulation of labour in our own dominions in regard to our Indian fellow-subjects, we shall be in a bad position when we endeavour to persuade foreign states to abolish the slave trade. I cannot admit this paradox, till the freedom I support shall be demonstrated to be slavery. In my opinion, the less we interfere with the natural freedom of our own labourers, the more effect will our arguments in favour of free labour have with other nations. I hope that we never shall be placed in the trying situation of having to choose between partially enslaving our own people in practice [by prohibiting the migration], or abandoning the cause of foreign slaves, in theory.[102]

Within the terms of liberal discourse, there is, of course, no way to demonstrate the paradox that the "natural freedom" Grant supported did not find embodiment in the contract but was an effect of it. In broader, Marxist terms, what cannot be admitted within the liberal understanding of freedom, and thence free labor, is that not only is the laborer free to sell her labor power; she is *compelled* to. What Grant would call the "abomination" of slavery had come to mean simply the *absence* of a contract.[103] Thus, the mere *presence* of a contract had come to serve as the preeminent, indeed, *only*, signifier and indicator of freedom. Freedom was merely the *ritual of consent* to a contract, severed from the material conditions it stipulated.[104] In the regime of Indian migration control, the ostentatious presence of the contract, Austin's quintessential performative, brought into being that which it supposedly embodied, not only consent, but freedom itself.[105] It thereby brought into being the consenting, autonomous subject of liberal–juridical power. (Simultaneously, as will be apparent in the next chapter, the migration of indentured Indian labor would cultivate another kind of subject, namely, the subject of disciplinary power.)

"The Majesty of British Legislation": The Modern State in a Colonial Field

Despite all the calumny heaped upon the Dickens Committee Report and despite its condemnation of every aspect of Indian migration, it had, in fact, also included the caveat of an elaborate system of state control to manage the migration, were it resumed. However, the way the Report had framed the overarching conclusion for a termination of the system raised, if inadvertently, the crucial question of colonialism: "We are thoroughly and intimately persuaded, from our knowledge of this country, of the working of judicial and police establishments[,] . . . of the character of the natives . . . that scarcely any human precaution would avail to prevent a repetition of abuses. . . . [In fact] *no system,* we are firmly convinced, would ever suffice completely to counteract the tricks and falsehoods that would be resorted to in India."[106]

Implicit, therefore, in the Committee's assessment was a tacit capitulation to the imminent failure of the civilizing mission and of the modern state in a colonial field. By casting the "character of the natives" as not amenable to reform, amendment, transformation, it put in question the very basis of the civilizing mission. Colonialism could operate only by simultaneously advancing the uncivilized status of the natives *and* their impending civility. To push either of these positions to their logical end would have undone the ideological justifications for colonial rule. For, if the native could not be civilized, then what was the purpose of colonialism? Alternatively, if the native acceded to the trappings of civilization, then what justified the continued colonial situation? Rescuing and reinscribing the narrative of the civilizing mission was, therefore, a constant, threatened project within colonialism.

State regulation of Indian migration became part of the civilizing mission integral to liberalism, understood not merely as a philosophical or theoretical tradition, but also as a material and historical relation. Or, to reverse Robert Young's statement, imperialism was not only a subject-constituting project but also, and more importantly, a territorial and economic project.[107] Demanding, we might say requiring, the continuation of Indian indentured labor were two crucial factors: the need for disciplined, cheap labor in the sugar plantation economies, linked directly to the sweet tooth that had developed in Britain, and the need to rescue "the rule of law," and thereby the project of colonialism, from the impotency the Dickens Committee Report attributed to it—by suggesting that *no system* could counteract the "character of the natives." Hence, in the Committee's judgment, liberal–juridical law and the system of British legislation had met its match in the native. This judgment is what the Mauritius Free Labour Association would grasp upon and deploy to the fullest extent, writing in their extended response:

It appertains to the Legislatures of Great Britain, of India, of this colony, to repel these disrespectful assertions of their incapacity, by showing that they are competent to frame and enact whatever laws the growing exigencies of society may require. . . . Let them do this, and then, whilst the frightful visions of the deluded philanthropist will completely vanish in the broad light of truth and justice, the angry discussions, the agitation of the public mind, excited by the sordid . . . will be hushed and disappear in the presence of the majesty of British legislation, maintaining and protecting the inestimable rights and liberties of British subjects.[108]

Among the signatories to this document, waxing eloquent on the majesty of British legislation, was Hollier Griffiths, the planter who, in the earliest days of Indian migration, had challenged *any* state intervention as an illegitimate exercise of sovereign power and as "so contrary to the liberty of the subject." We can see how the discursive field had shifted between 1835, when Griffiths objected entirely to legislative intervention, and 1840, when he signed a document extolling the virtues of such intervention. This was a discursive shift whose burden was borne by the "character of the natives" and the tendency of the African free laborer to demand "monopoly" wages, which, happily, permitted the resumption of Indian migration.[109] Meanwhile, those in the metropole continued to discover the many uses of cheap sugar.[110]

The tone of the response from the Mauritius Free Labour Association might be dismissed as the hyperbole of planters eager to rescue their tarnished credentials as ex-slave owners in a world where the predominant term of the politico–economic lexicon was "freedom." However, even the measured tones reflecting "median positions," in recommendations such as those coming from James Grant or Charles Anderson, advanced the same solution. Grant summed up the misguided position of the Dickens Committee Report in the following way: "The argument we are recommended to hold to the emancipated slave colonies to my mind is this: formerly you had nothing but slave labour. Slave labour is an abomination, therefore we shall now do all we can to prevent your having free labour."[111] In his view, the "evils which attended the export of Indian labourers" were of a "casual" nature, easily prevented "by good regulation."[112] The *pièce de résistance* of the regulations Grant offered to cure the system of its "casual evils" was the contract, though now it was to be authorized by a specially appointed Protector of Indian Emigrants who "should have no [personal and vested] interest in the matter." In many respects, the details of Grant's proposal were similar to those offered by the Dickens Committee, if less minute and elaborate. The key distinction, however, was that he upheld the view that the system could, in fact, be managed under the supervision of the British legislature and via the appointment of "impartial" state functionaries such as a Protector of Indian Emigrants. Grant, in other words, explicitly advanced the civilizing mission.

Charles Anderson, acting now on behalf of the Mauritius Free Labour Association in London, would arrive at similar conclusions. Writing to Lord John Russell at the Colonial Office, Anderson echoed Dowson and portrayed the planters in Mauritius in dire circumstances and on the verge of "irretrievable ruin" with Indian migration as the only solution.[113] And, like Grant, or indeed like the Dickens Committee Report, "the fundamental principle" of his proposal was government supervision of indentured migration, in both India and Mauritius.[114] This, opined Anderson, would give "a public character to the general undertaking, which could not fail to prove highly advantageous to its proper direction."[115] Anderson believed that such a measure would convince "even the most skeptical opponents . . . that the emigration of Indian labourers to Mauritius, on the amended system, would have the most beneficial effect on the state and condition of the emigrants, and [moreover] it is only by their immediate introduction that the cultivation of sugar-cane can be continued."[116] The crux of "the amended system" was state regulation. As Thomas Hugon, one of the members of the inquiry committee in Mauritius (the same committee with which Anderson had taken issue), would bluntly frame it: "Totally different principles . . . form the basis of the new system. . . . Under the first, the interference of Government was studiously avoided, and even opposed; here it is brought in wherever it can prove efficacious in the prevention of abuses."[117] While the tenor of proposals from those such as Grant, Anderson, or Hugon was markedly different from the exaggerated tones of the planters, their premise was the same: a belief in the "majesty of British legislation, maintaining and protecting the inestimable rights and liberties of British subjects." The authors of the Dickens Committee Report, in fact, also held this belief, though it was submerged in the flurry of agitated responses. Such beliefs were not consolidated in a vacuum but emerged, instead, from material exigencies—here the imperative to enable the continuation of indentured migration in the interest of the sugar plantation economy.

Gone were the days when state regulation of migration was challenged as an infringement on the "liberty of the subject." The activation and circulation of a colonial/modern discourse on freedom enabled the boundary between the principle of sovereignty and the principle of the liberty of the subject to be drawn in favor of the principle of sovereignty. It would be the rule of law under a strong sovereign that, as Macpherson suggests, becomes particularly crucial in moments of transition that would ensure the "inestimable rights and liberties of British subjects." This sovereign would dictate, per Hobbes, "in what manner, all kinds of contract between Subjects . . . are to bee made; and by what words, and signes they shall be understood for valid." Here, then, is a portion of the "simple" regulations Grant offered to ensure that the contract embodied the consent and free will of the native; however—and this is crucial—his minute made sparse reference to equality in exchange, or to the fairness of the terms of the contract:

> At his office [the Protector, of European origin and descent] might examine leisurely one by one, and without the presence of any party interested in the system, every labourer who should come to him professing a desire to embark. After making sure that the labourer fully understands what he undertakes, and that he is acting voluntarily, the Protector might make out a certificate containing a description of the holder's person. A registry of all certificates . . . might be kept. The system of passing [i.e., examining] coolies in batches might be forbidden. It might be the Protector's duty to see that no force or show of force . . . be allowed. . . . The certificate might . . . be a sufficient warrant to any master of a vessel who had obtained authority from the Protector . . . to receive the holder [of a certificate] on board. The protector might refuse to grant such authority, unless convinced that the master and officers are . . . suited for such a duty as the conveyance of Indian labourers. Before the ship breaks ground, the Protector might proceed on board to grant a final permit to the master to sail with such labourers as [those who] then and there produce their certificates, and still testify to their desire to proceed.[118]

It is now possible to more fully appreciate all that is at work in the 1842 executive order of Lord Stanley, Secretary of State for the Colonies, which reopened indenture, and with which I began this chapter. It is worth recalling:

The abolition of slavery has rendered the British colonies the scene of an experiment, whether the staple products of tropical countries can be raised as effectually and as advantageously by the labour of freemen, as by that of slaves. To bring that momentous question to a fair trial, it is requisite that no unnecessary discouragement should be given to the introduction of free labourers into our colonies. So far as it may be inevitable to obstruct such immigration by taking effective securities that the immigrants shall be, in the fullest sense of the term, free agents,—that obstruction may be justified, but no further.

The "obstruction" that was "justified" was the (state-supervised) contract, which would serve as an "effective security" that the immigrants were in "the fullest sense of the term, free agents." Thus Stanley's "momentous question" was brought to a "fair trial" and it, indeed, proved that "the staple products of tropical countries can be raised as effectually and as advantageously by the labour of freemen, as by that of slaves." In the process, however, "the fullest sense of the term free agent" required only the presence of a contract whose validity began and ended with consent, dispensing with fairness and with equality in exchange. Perhaps, then, the paradigmatic site for the separation of consent from the notion of equality in exchange, which characterizes the nineteenth-century reformulation of the contract, is to be found not within the metropolitan heartland, but within the peripheral sites examined here.

Conclusion

In its bare outlines, the migration of Indian indentured labor simply constituted a source of labor necessary for the continued and expanded cultivation of sugarcane in the ex-slave plantation economies. This economic imperative, however, had effects and repercussions that were infinitely more complex. Indeed, the *form* of the migration and the debates it provoked raise important questions in multiple domains. Here I have explored, in particular, how the seemingly disjointed events of the abolition of slavery in Mauritius and the Caribbean, the colonial situation in India, and the ascendance and consolidation of key tenets of political economy in Britain, as also among its colonial functionaries, are linked phenomena. An attention to these linkages enables, I believe, an enriched understanding both of the centrality of consent to nineteenth-century contract law and of the reconfigurations attendant on the making of modern state sovereignty.

It is, I think, arbitrary and misleading to attempt to understand the transformations in nineteenth-century contract law without taking account of abolition, be it at the literal sites of the ex-slave plantation economies or at the multiple sites, ranging from India to France or Britain, that were articulated to the event in *direct*, yet complex, ways. Further, since we witness a contemporaneous emergence of the centrality of consent to the contract in such multiple sites as Mauritius, India, and Britain, we are forced to reevaluate how understandings of paternalism were implicated and imbricated within this transformation. If contract theorists have attributed the rise of consent in the nineteenth century as working against the paternalism that was thought to underwrite the concept of equality in exchange, hitherto required for a valid contract, in the colonies, this paternalism worked differently. As described above, state regulation of indentured Indian migration was justified in terms of the "necessary ignorance" of the natives. This "ignorance," combined with the propensity of native recruiters to deceive and defraud potential migrants, thus demanded that the colonial state itself became the arbiter of the terms of indentured contracts. Thus the *paternalism* of the state enabled it to set the terms of (unequal) contracts and simultaneously required each migrant to "freely" consent to these terms. This act of consent, before the *European* Protector of Emigrants, was now held as sufficient to guarantee the validity of a free contract. The rule of colonial difference was, in this way, at the heart of the making of the colonial/modern state.

State intervention and supervision of Indian indentured "free" migration is also important in terms of the debates it occasioned on the limits and authority of state sovereignty. In its initial stages, these conflicts were resolved, once again, by an appeal to the exceptionality of the colonies. However, by about 1840, these debates were almost completely submerged and, indeed, reversed. Whereas Hollier Griffiths, in 1835, had pointed to the perils of state intervention, in 1840, he would point to the promise of such intervention. Lord Stanley's order resuming indentured migration to Mauritius is dated January 22, 1842. In 1844, he reopened indentured migration to British Guiana. Over the course of the nineteenth century, Indian indentured migration would be authorized (and, at times, suspended) for more than twenty countries on four continents. In each case, state supervision of the migration was authorized as an *exception* to the general principle of free migration, while also coding the migration as "free." The paradoxical situation produced by this historical relation, where the state regulated "free" migration precisely in order to ensure that it was "free," cannot be resolved. It is best understood in terms of the historical circumstances—political, economic, legal, and ideological—that produced these paradoxical solutions, and complex, irreconcilable new discourses of freedom in the making of the world.

At the same time, we should note that state intervention in regulating indenture has the distinctive feature of operating within a *logic of facilitation* and not within a *logic of restriction*. The latter is often thought to be the defining feature of state regulation of migration. Such logics of restriction, which would dispense entirely with the principle of free movement in the name of a specifically *national* sovereignty, would emerge as the ad hoc resolution to new historical circumstances. These later developments, while generating a novel understanding of "nation" and "national sovereignty," would, among other things, have to navigate the paradoxical understandings of freedom and the specific legal and legislative definition of an "emigrant," produced from the conjuncture between slavery, abolition, and indenture. I turn to some of these developments in chapters 3 and 4, which will demonstrate the colonial genealogy of the articulation between "nation" and migration and its enduring institutional legacies. But first, in the next chapter, I seek to provide an overarching analysis of the formation of a massive state bureaucracy to regulate Indian indentured migration. This analysis examines some of the minutiae that characterize the migration regimes of the modern state apparatus that, I argue, are usefully apprehended in terms of Foucault's notion of disciplinary power.

CHAPTER 2

Disciplinary Power and the Colonial State

The Bureaucracy of Migration Control

As we saw in the previous chapter, the Dickens Committee Report (or the "majority report"), submitted by three members of the inquiry committee appointed in Calcutta (Kolkata) to assess the system of Indian indentured migration, condemned most every aspect of the system and deemed it unsalvageable. The dissent of committee member James Grant to the negative assessment of the majority report played no small part in ensuring a resumption of the migration. In Grant's view, the "evils" attendant on indenture were of a "casual nature," easily prevented "by good regulation."[1] Outlining the contours of such regulation, Grant was primarily concerned to systematize state oversight for ensuring that the contractual arrangement embodied consent. His recommendations suggested a series of "simple" processes to facilitate the legal corroboration of consent, such as: (1) selection of an "impartial" Protector of Emigrants;[2] (2) his "leisurely" examination of the native emigrant; (3) the absence of parties "interested" in the matter; (4) procedures for documenting identity; (5) issuance of a certificate that testified, simultaneously, to identity and to consent; (6) formulation of a bureaucratic practice to manage the official record of the certificates issued; (7) evaluation of the "suitability" of the master and the crew; and (8) a process of final verification that those on board a ship had their certificates and reiterated their "desire to proceed."[3]

Dormant in the language of these interlinked steps is vast ambiguity. What, exactly, was an adequate, if leisurely, examination of the emigrant? What, precisely, defined an "interest" in the matter of emigration? What details, specifically, constituted what Grant termed a "description of the holder's person"? How was identity to be ascertained and documented? What elements constituted the "suitability" of the master and the crew? How was his suggested "registry of all certificates" to be generated and retained? Such questions, among many others, would become persistent and produce an ever-increasing refinement in the legislation, rules, and regulations. However, despite the aid of law and of rules, achieving the dream of uncontested exactitude and unsullied precision would remain elusive.[4]

Regardless of the many unresolved ambiguities inhabiting Grant's proposal, the legislation and rules that were drafted followed his recommendations and retained the minimalist nature that heretofore had characterized migration regulations—Act XV, passed on December 2, 1842, following Lord Stanley's decision to reopen Indian migration to Mauritius—was a mere seven pages.[5] These seven pages encompassed the regulations to be followed in India—at the ports of Calcutta, Bombay (Mumbai), and Madras (Chennai)—as well as in Mauritius and summarily covered what appeared to be the important issues: the appointment of a salaried em/immigration official at both sites, the necessity of a "certificate" or "pass" corroborating the emigrant's "willingness and desire to work for hire in the said colony of Mauritius," the stipulation that the passenger ships not be overcrowded, the unambiguous arithmetic that for the purposes of "the computation of the number of passengers" every two children under the age of ten years would count as one adult, and the requirement that there be sufficient "good and wholesome provisions" for the emigrants' consumption, distinct from those provided for the crew.[6] In addition, the Act specified that an emigrant would not "be capable of entering into any contract for service . . . until he shall have been at least 48 hours on shore."[7] This caveat seems to have entered the regulations in an acknowledgment of the point raised by the Dickens Committee Report, discussed in the previous chapter, concerning the unfairness of having emigrants sign contracts in India prior to their departure, and thus in ignorance of the conditions of the labor market in Mauritius. This 48-hour caveat would be short-lived, as would other aspects of the brief Act.

Barely a year after its ratification, it was deemed inefficient for the Government of Mauritius to fund the salaries of emigration agents at each of the ports of Calcutta, Bombay, and Madras.[8] As a result, and with fresh legislation, Indian emigration to Mauritius was restricted to the port of Calcutta by Act XXI of 1843.[9] Following a significant correspondence, one year later, Indian emigration to Jamaica, Trinidad, and British Guiana (Guyana) was authorized via Act XXI of 1844.[10] This Act largely replicated Act XV of 1842, with the crucial distinction that emigrants contracted to labor in India. Thus was eliminated the possibility, in these colonies, that emigrants have at least 48 hours at their destinations prior to making contracts. Lord Stanley, a staunch supporter of redistributing labor within the empire, took issue with certain aspects of this Act and wanted to extend the Passengers Act in force in England to India. This, however, did not come to pass; however, on December 20, 1845, Act XXV of 1845 was ratified to address some of his objections. In sum, directly following the resumption of indentured migration, fresh legislation regarding Indian migration was ratified almost every year. Put in motion thereby was an unending series of legislation, rules, regulations, annual reports, quarterly returns, routine correspondence, confidential correspondence, routine memos, secret memos, routine inquiries, special inquiries, and the like, all attempting to micro-manage Indian migration for the next seventy-five years.

In this chapter, I turn our attention to these endless rules and regulations and the accompanying formation of a massive bureaucracy for regulating the "free" migration of indentured labor. As I have noted in the introduction, in the continual process of accretion that produced the gargantuan machinery of minute techniques and technologies to manage every aspect of Indian indentured emigration, the colonial state generated and deployed a range of uniquely modern mechanisms characteristic of what Foucault calls disciplinary power. If sovereign power is discrete, is localized, and relies on force, with law as its overarching template, disciplinary power is uninterrupted, is continuous, works through minute rules, and is a "political anatomy of detail."[11] It seeks to monitor every activity, to organize all behavior, to surveil every individual, and to infiltrate all relations "right down into the depths of society."[12] Discipline is a modality of power that is mobile and transferable and can invest itself in all manner of apparatuses and institutions—be they the military, the educational system, the hospital, or—most famously—the prison. As we shall see here, it also inserted itself into all aspects of the bureaucracy and apparatus for managing Indian migration. Keeping in view the distinction between categories of practice and categories of analysis (discussed in the introduction), this chapter aims to make *analytical* sense of a vast and complex bureaucracy, even as it seeks to provide a textured account of the "political anatomy of detail" that characterized what was, simultaneously, a chaotic and messy web of legislation, rules, and regulation. I turn, first, to making analytical sense of the chaotic muddle of regulations concerning indentured Indian migration, before addressing what was an emblematic instance of bureaucratic rationalization and standardization.[13]

The Migration Apparatus: Part I: Inconsistency, Expedient Logics, and the Coherence of Incoherence

As I have noted, state control of Indian migration, of labor under contract, operated on the logic of exception. Special legislation was required to authorize indentured migration for each colony approved as a destination site. Thus, in the space of some twenty years, between 1842 and 1864, there were nineteen different Emigration Acts in force in India. The provisions of these different Acts were not in consonance with each other: the laws and regulations varied from colony to colony and even between ports of embarkation. The rules in force at Madras for migration to Mauritius differed from the rules applicable at Calcutta; both sets of rules, moreover, differed from the rules relating to the migration from either port to Trinidad or Jamaica. Thus, on the one hand, the system—if it can be called that—was inordinately messy and inconsistent such that it is impossible to make any overarching statement regarding points of detail. This is true even for crucial issues where one might expect regularity, such as the length of the contracts, the rules governing return passages, wages (which differed even for the same colony), whether the contracts would be made in India or at the destination, or the quantity of food and water to be supplied on voyages.

For instance, as noted above, the legislation of 1842, which resumed migration to Mauritius, stipulated that only contracts made by migrants after 48 hours in Mauritius would be deemed valid. This stipulation was overlooked in the legislation enacted just two years later for Trinidad, Jamaica, and British Guiana. Whereas, in the initial phase, the destination colonies were responsible for guaranteeing migrants a free return passage after five years in the colony, this would soon change and take a number of different forms. These changes ranged from simply eliminating any responsibility on part of the destination colonies, to withholding a portion of emigrants' wages toward a fund to pay for the return journey, to stipulating a free passage after ten years' "industrial residence" in the colony, to offering a free passage after ten years in the colony (with five under "industrial residence"), to the option of a "bounty" in lieu of foregoing the passage, to the grant of Crown lands in lieu of the return passage.[14] Similar disharmony prevailed on lesser points such as the dietary scale in force at the embarkation depots and for the voyages. The discrepancies included the fact that, inexplicably, according to rules in force at Calcutta, emigrants to Mauritius "were given 4 oz. more rice per diem than emigrants to the West Indies."[15] At Madras, despite legislation mandating that ships carry seven gallons of water per week for each emigrant, the rules promulgated required the provision of only five gallons per week. Though strictly illegal, apparently the legislative requirement had been "overlooked when sanctioning the rules."[16] Hence, in a certain way what perhaps characterizes the system put in place to manage Indian migration is its lack of consistency and coherence. In the letter accompanying his landmark 1873 report, "A Note on Emigration from India," Mr. Geoghegan would make the same point, writing, in frustration, that his review of some thirty years of Indian migration law and policy served "conspicuously to illustrate the absence of any consistent line of conduct in dealing with questions of emigration."[17]

Mr. Geoghegan was not the first contemporary to note with dismay the disharmony and incoherence of the rules surrounding Indian emigration. Some ten years earlier, in 1864, the confusing state of affairs had been remarked on by none other than Mr. Henry Maine (later Sir Henry Maine). Maine arrived in India in 1863, upon being offered the important position of Legal Member of the Council of India, directly following the publication, in 1861, of *Ancient Law*, his landmark treatise.[18] The thesis put forth in *Ancient Law*, and forever associated with Maine's name, is that societies moved from being organized by rules of status to rules governed by freedom of contract; in a nutshell, law and society evolved "from status to contract." If Maine would put forth this thesis in academic form in 1861, its "truth" was already being elaborated and put into practice in a range of settings by British bureaucrats, officials, and politicians concerned with governing the colonial world. As we have seen in the previous chapter, the management of Indian migration was one site where the shift to contract was explicitly mandated.

When Maine arrived in India in 1863, the contract was already in place to regulate Indian migration. His contribution was to attempt to rationalize and systematize the haphazard series of laws and further refine the rules of consent governing the contract. Though many scholars have noted that Maine broke with the principles of utilitarian thought, with respect to regulating Indian migration, it was Maine-as-Utilitarian who was in evidence.[19] Under his supervision, a number of important changes were effected that together formed Act XIII of 1864. These changes included consolidating the numerous Acts in force into one new Act, thus eliminating the unwieldy system of introducing fresh legislation for each colony approved for emigration. Instead, Maine helped craft general legislation that could be applied to each new case. Along with simplifying the process by which new countries were authorized as legal sites for emigration, this aimed to make the system consistent over a range of colonies and intended, especially, to remove the discrepancies "between the system of emigration to French colonies and the system under which emigration takes place to dependencies of the British Empire and certain other localities."[20] Despite Maine's efforts, however, the regulation of Indian migration remained resistant to regularity and systematization. The machinery never fully stabilized and, in the vast documentation it produced, there was always a crisis unfolding: either the rules and regulations in place were insufficient to deal with the circumstances or the violation of rules necessitated a series of inquiries resulting in ever more minute rules. Consequently, the system was in constant flux and exceedingly unwieldy as it sought to incorporate the bewildering complexity of human existence into a set of rules.

In contrast and at cross purposes to the inconsistencies and discrepancies in the system was the willingness of officials to make searching inquiries, to conduct extensive transcontinental correspondence on minute points, to draft rules to address every circumstance, and to frame legislation that sought to address every calamity and refine every oversight. As a result, what emerged was not a dearth of rules, regulations, legislation, information, or documentation for managing Indian migration, but a plethora of them. However, in spite of the inconsistency and incoherence of the laws and regulations, the scale of their operations, and the sheer surfeit of material documentation, it is possible to identify certain regularities of form, overarching logics, grammars of control that, despite Mr. Geoghegan's frustrated assessment, did, in fact, lend coherence and produce a *system* or *regime* of state control of Indian migration.

The Migration Apparatus: Part II:
Disciplinary Power and the Making of a Bureaucracy

Among the logics of this system, one can identify, in schematic fashion, a number of dimensions of the migration control system: (1) the twin logics of surveillance and examination operating in various modes; (2) the emergence of certain figures serving as nodal points within the complex network of emigration control; (3) the delineation and emergence of certain spaces as key sites of intervention; (4) an ever-intensifying concern with the body in terms of health, disease, and mortality; and (5) as a complement to surveillance and examination, the increasing systematization of each aspect of the migration through the call for meticulous record keeping. These logics often worked in conjunction with each other and there are additional logics one can discern and identify in the ever-expanding universe of Indian emigration control. I will address some of these additional logics in subsequent chapters; here, I restrict my discussion to these five elements.

The Logics of Examination and Surveillance

The regulation of Indian migration was attended by a proliferation of different modes of surveillance ranging from the examination of the emigrant to the supervision of every official, up and down the grid of bureaucracy that came into being. This proliferation followed, in textbook fashion, what Foucault calls a system of "hierarchized, continuous and functional surveillance," a crucial element of disciplinary power.[21] This form of surveillance, while utilizing individuals, is dependent on a grid. It works by way of "a network of relations from top to bottom, but also to a certain extent from bottom to top and laterally; this network 'holds' the whole together and traverses it in its entirety with effects of power that derive from one another: supervisors, perpetually supervised."[22] The regulations surrounding Indian migration sought to bring to perfection the possibility of this continuous grid of surveillance and supervision.

Beginning with a cursory, yet racially saturated, examination of the Indian emigrant by the European Protector of Emigrants, modalities of surveillance and examination would soon expand. The unceasing complaints of fraud and force that dogged Indian migration propelled Henry Maine to rework the authority and spatial coordinates for ascertaining the migrant's consent. The reworked scheme required that each emigrant be examined and registered by the magistrate in the district where he or she was recruited. Whereas, previously, this task belonged to the Protector of Emigrants at the port of embarkation, Act XIII of 1864 introduced into the bureaucracy another functionary: the District Magistrate. Henceforth, the Protector of Emigrants positioned at the ports would "no longer conduct the inquiry as to the free will of the emigrant," except in those cases where the emigrants were recruited at the embarkation ports.[23] It was now the responsibility of the local magistrate in the recruiting district to "interrogate the recruit as to his comprehension of the engagement and willingness to fulfill it."[24] The District Magistrate was required to forward to the Protector of Emigrants, at the relevant port of embarkation, a list of the emigrants who had consented to emigrate. Though the Act relieved the Protector of the task of inquiring into the thorny matter of consent, it also outlined, for the first time, "a legal description and definition of the duties of the Protector of Emigrants."[25] Thus the Protector, charged with a host of other tasks, constituted an important nodal point within the apparatus of migration control.

Interrogation of potential emigrants was but one site in the vast network of examinations, supervision, and surveillance in the apparatus of migration control: over the years, as the emigration bureaucracy expanded and its rules congealed, no official was authorized to conduct any action without supervision and surveillance, either direct or indirect. Following the passage of Act XV of 1842, low-level functionaries such as recruiters were initially selected and appointed by the Emigration Agent (who, in the early years, also doubled as the Protector of Emigrants) for Mauritius.[26] These recruiters were urged to avoid fraud and force in recruitment and asked (but not required) to register the emigrants with the local District Magistrate or "nearest European functionary."[27] It appears that despite the extension of Indian migration to a range of colonies, these recommendations were rarely followed. Complaints of deception and exploitation continued unabated. Thus, by 1864, the recruiters would no longer be selected and appointed by the Emigration Agents representing the destination colonies but were to be licensed by the Protector of Emigrants; and every such license, valid for one year, would have to be counter-signed by the District Magistrate of the locality of recruitment.[28] Such precautions, too, seemed to offer little relief from the avalanche of complaints. As was the case at the very inception of the system in the 1830s, deceptions and instances of fraud were attributed to the questionable and unsavory character of some recruiters.[29] Consequently, some twenty years later, in 1883, it became necessary to include in the set of rules an entire section devoted to "Recruitment," and, among the several forms required for licensing a recruiter, was one titled "Form of Certificate of Intending Recruiter's Character."[30] A *record* of the examination, surveillance, and licensing of the Recruiter would become in this way ever more minute.

Surveillance was not merely the lot of the functionary at the bottom of the hierarchy. A similar pattern of increased scrutiny, multiplying forms, and incessant record keeping attended the functions of every official inserted into every level of the bureaucratic hierarchy. The requirement of constant reports and "returns" was one of the most frequent methods of exercising surveillance. Witness, for instance, the 1883 rules outlining topics to be covered in annual reports submitted by key emigration officers:

The submission of annual reports by emigration officers shall be regulated as follows:—

a. The Magistrate of every district of recruitment shall forward to the Protector of Emigrants concerned as early as possible, and not later than two months after the close of every official year, his report of that year, reviewing the conduct of recruitment for the colonies within his jurisdiction, with special reference to the following points:—The condition of the places of accommodation; the irregularities, if any, committed, and the punishment of the persons concerned; the prevalence of any epidemic disease; and the local circumstances, such as the state of the labour market, and the price of food-grains, &c., which either favoured or hindered emigration.

b. Every Emigration Agent shall similarly forward his report, reviewing the operations of the past year, with special reference to the results of recruitment on his behalf; the terms of engagement of labourers; the disposal of emigrants accommodated in depot; the sickness and mortality in depot; the vessels dispatched; the number of returned emigrants; and the amount of their savings.

c. Every Medical Inspector of Immigrants shall similarly forward his report, treating on the following points, namely—the condition of the depots; the state of health of the emigrants accommodated therein; the sufficiency of hospital and other medical arrangements; the cause of any epidemic disease; and the precautionary measures adopted against its spread.

d. Every Protector of Emigrants shall, not later than four months after the close of every official year submit to the local Government his report for that year. The report, while embracing information on all points of importance noticed in the annual reports of other emigration officers received by him, shall review the history of emigration during the year, noticing all points of importance; and shall furnish such other particulars as may from time to time be required by the local Government.[31]

The information required in the report of the Protector of Emigrants seems to allow meaningful latitude for discretion, yet his duties had been systematized and extensively listed in the Rules accompanying Act XIII of 1864.[32] In short, processes of examination and surveillance pervaded every aspect of the apparatus of Indian migration control and, with each passing year, became simultaneously more minute and cumbersome.

Official Networks and Positions of Control

We have seen above how the District Magistrate and the Recruiter came to occupy significant positions in the grid that formed the bureaucracy of Indian migration. Unlike the District Magistrate, who was merely assigned a new set of tasks, the position of the Recruiter, like that of the Protector of Emigrants, emerged as a consequence of the specific constraints and logics that organized migration control. Further, some of these positions, such as Recruiter or Plantation Manager in the different destination colonies, had lineages that could easily be traced to the slave trade and slave plantation economies. Indeed the lineage was often so evident that the problem confronting the colonial administration was how to sever these connections and invest the various positions with new significations.[33] The primary method used in distancing Indian migration from such associations was the constant invocation of the "wellbeing" of the emigrant, that, in keeping with utilitarian principles, was to be achieved by drafting further laws, rules, and regulations.

Increasing regulation of Indian migration was a part of the nineteenth-century expansion of the state apparatus, more generally, and the post-1857 expansion of the colonial state in India, more specifically. Hence, just as the mechanisms of supervision and surveillance intensified over the years, the number of officials, semi-officials, and gradations in the hierarchy of migration control also increased in stratification and density. Within the complex network of officials that emerged, certain figures served as nodal points, especially important transfer points of authority and control. As it turns out, the emigrant—particularly the male emigrant—was a minor figure in this structure. In the early twentieth century, it would be emergence of another figure, the Indian-woman-made-dishonorable-through-migration, mobilized by Indian nationalist discourse and a mass movement for the abolition of indenture, who would help upset the entire grid that held the system together, and thus facilitate its demise.[34] Be that as it may, as the system was taking shape, the chief figures who served as nodal points were the Recruiter, the Emigration Agent, the Protector of Emigrants (in India), the Protector of Immigrants (in the different destination colonies), the Medical Officers (in India, on board the ship, and in the destination colonies), the Sirdar (native headman/supervisor on the ship or the plantation), and the Plantation Manager (plantation owners were often absent). The position each functionary occupied in the hierarchical grid of the migration bureaucracy naturally determined the degree of authority and control he exercised, with the Protectors located at the apex of the pyramid. However, while the Protectors, the Emigration Agent, and the Medical Officers might be said to constitute the top tier of the institutional structure, the Recruiter, the Sirdar, and the Plantation Manager, located at lower tiers, were especially crucial positions. Given their proximity to the emigrants, they were an easy target for blame from those higher up the hierarchy, and their actions and (mis)management generated a constant litany of complaints from emigrants and other parties. This combination was frequently responsible for the unending series of refinements that expanded the rules and regulations.

Whereas certain positions, such as the District Magistrate or the Plantation Manager, antedated Indian migration, the tasks and responsibilities invested in these positions were partially transformed as they were inserted into an emergent regime of migration control. Simultaneously, and alongside these crucial transformations, the very existence of these positions already presumed certain practices and evoked dense and complex histories. For instance, when, in the late nineteenth century, Fiji was approved as a colony for Indian emigration for the purposes of sugar cultivation, the model of the plantation along with the position of the Plantation Manager, saturated with the multiple legacies of slavery and of indenture, was transplanted to (and transformed at) a site that had not been subjected to chattel slavery.[35] The position of other functionaries in the grid can be understood as more "organic," emerging as a result of various context-specific exigencies such as the chain of events that led, in 1862, to the appointment of a Medical Officer to oversee the depots at the ports of embarkation, and to the employment of professional cooks, "one for every 100 Emigrants," on board ships. High mortality on board emigrant ships caused the Court of Directors, in 1857, to initiate an inquiry into causes.[36] In 1859, with no "perceptible diminution taking place on the heavy rates of mortality" and with the transfer of power to the Crown following the Indian Revolt of 1857, the Lieutenant Governor of Bengal established a committee to inquire into "every point connected with the deportation of Coolies."[37] The Medical Officer and the cooks were appointed at the recommendation of this committee and soon became part of the regulatory apparatus.[38]

In this way, the web of bureaucracy that formed to manage Indian migration utilized existing bureaucracies and officials even as it created a host of new positions that, over the years, included Interpreters, Medical Officers, Assistant Surgeons on ships (who could be "natives"), Sweepers, Cooks, Sirdars, Emigration Agents, District Magistrates, Protectors of Emigrants, Protectors of Immigrants, Plantation Managers, Recruiters, Sub-Agents, Police Officers, and all manner of special officials appointed to conduct innumerable special inquiries into the system. Even as each of these positions ostensibly emerged to fulfill certain tasks, the tasks themselves needed specification in meticulous detail, with responsibilities clearly outlined and the chain of command explicitly stated. The networks of surveillance and number of elements in the structure thus became part of an increasingly densely populated and ever-expanding web.[39]

Temporalizing Mobility, Spatializing Control

In our current moment, perhaps the most significant spatial locations of migration control are iterations of the border, whether or not this constitutes the literal border demarcating nation-states (for example, international airports).[40] Within the regime of Indian migration control, the sites that crystallized as particularly significant for surveillance, management, and intervention were the embarkation depot, the ship, and the plantation. On occasion, other sites would become the target of intense scrutiny—the office of the District Magistrate in India, or a particular recruiting district, or the path of the journey from interior districts to the coast, or the route, in the destination colonies, from a given plantation to the magistrate's office where one could lodge a complaint.[41] Though subject to regulation, these sites did not congeal into locales that became part of the legalistic and disciplinary apparatus of migration control with quite the same intensity as the embarkation depot, the ship, and the plantation. The ship and the plantation had been, in different degrees, the object of scrutiny from the very beginning (for instance, in requirements concerning the overcrowding of ships and restrictions on the number of passengers allowed on board). The embarkation depot, however, was consolidated as a unique site worthy of special attention and regulation in Act XIII of 1864, under the guidance of Henry Maine. The aim of the regulations Maine helped devise was to ensure, as he put it, that the depots be "liable to constant inspection" by the Protector and the newly instituted Medical Officer.[42]

The appointment of a Medical Inspector combined with the codification of regulations did, in fact, ensure that every aspect of the depot was subject to scrutiny. Indeed, having both a permanent official and a permanent depot, as was the case with the depot for Mauritius at Calcutta,[43] tended, as Surgeon Partridge, the temporary Medical Inspector, observed, "to render easy the carrying out of a system of steady supervision in little things."[44] Prior to legislative action mandating the position of Medical Inspector, Partridge was appointed to the position in something of an experimental capacity. His detailed report on the Calcutta depots was, in part, responsible for formalizing the position and making the depots a spatially distinct site of intervention. The report covered a variety of topics: the physical location of each depot, its proximity to the river, the vegetation surrounding the depot, the drainage of the soil, the precise construction of the privies, the sleeping arrangements of the emigrants, and the abilities and dispositions of the Emigration Agents and of the Depot Medical Officers in charge. In the future, these elements and many more were minutely investigated and detailed rules on each were prescribed. Driven by the logic of hygiene and sanitation, privies and latrines were of special concern, both at the depots and on the ships. Partridge's report, for instance, contained extensive descriptions and comparative evaluation of the privies at the different depots. In his view, the Mauritius depot, though very well run, had privies that were "faulty in construction, unnecessarily complicated and not very well built"; the Jamaica depot was extremely well located and had privies "constructed on the American method on piers stretching into the stream, and reaching beyond the low water mark"; the Trinidad depot was "badly placed," at some distance from the river, and thus "the night soil has to be carried away by hand"; the Demerara (British Guiana) depot was not only inefficiently run but its location was "totally unfitted for the purposes of an Emigration Depot." The site, wrote Partridge, "is in wet weather a swamp, is surrounded by stagnant ditches, and under present circumstances is most certainly beyond the reach of any practicable improvement by local drainage."[45] Unsurprisingly, reports such as these would cause the formulation of detailed rules for the construction of privies at the depots and on the ships. Consider, for instance, a portion of the rules laid out in 1871 for the privies on emigrant ships:

> Water closets must be firmly fitted inside the vessel in two blocks, divided into compartments, the forward one (usually opposite the fore hatch) for the men, and the after block, as far aft as it is conveniently possible, for the women and children. The number of compartments and seats, according to scale in the Protector's office, will depend upon the number of statute adults to be conveyed. The closets must be well built, securely covered on their tops with tarred or painted canvas, and every compartment must be fitted with strongly hinged doors capable of being fastened on both sides; each compartment should not exceed 22 inches in breadth, nor should the seats in them be more than 22 inches square, though the *length* of the *compartment* should exceed its breadth by a few inches in front of each seat for convenience in entering the closet. The foot-boards on either side for the seats must be about 7 inches wide and 8 inches apart, so as to leave a central space of that size to communicate with the closet floor and shoot, which (lined with sheet iron) should be as far as possible free from all angles or projections likely to obstruct. The shoots must carry the soil well clear of the ship's side. If considered dangerously large for young children, the ports or scuttles through which the shoots lead must be guarded with

horizontal iron or hard wood bars to prevent accidents. A short strong hand-rail must be fitted inside the closet, and in a convenient position to hold on by. Those parts of the wood-work and shoots exposed to any source of contamination must be thoroughly and continuously lined with stout sheet iron, to prevent the wood from absorbing impurities, otherwise the floor and shoots must be constructed entirely of sheet iron or zinc of sufficient thickness. If the water closets are unavoidably high, convenient steps must be provided.

As represented by the model at the office of the Protector of Emigrants the closets should be fitted with a simple recess for the reception of a well-secured safety lantern, so that one light shall illuminate many compartments at the same time. . . .

Strong suitable screws and not nails must be used throughout in fastening the fittings at those points where they are subject to any strain likely to drag their parts asunder. The Master's special attention is drawn to this, as all the fittings must be well put together, and all hinges, bolts, clamps, sheet iron pipings or coverings, cowlheads, and every other description of fittings in which metal is used, must be thoroughly strong, well made, and well fastened.[46]

Instructions with this degree of detail were not unusual; rather, it became routine for rules pertaining to every aspect of the architecture, construction, and daily life at the depots, on the ships, and at the plantations to incorporate an excruciating degree of detail, which permitted extensive micromanagement as well as innumerable possibilities for infractions. At each site, there were explicit instructions for the separation of married couples from unmarried emigrants and of men from women. *Parda nashin* (veiled) women were prohibited from entering depots; the ships' crews were prohibited from having "any intercourse or to interfere, with the emigrants"; no female emigrant, in particular, was permitted to "be a servant to the Surgeon, cabin passengers, or officers of the ship."[47] Each site required a Medical Officer to oversee the health of the emigrants; the sick required separation from the healthy, necessitating that a section be cordoned off to serve as a hospital.[48]

Over time, regulations dictated how the emigrants were to spend the day at the depot and on the ship, mandating when they slept and when they rose, how and when they bathed, the process for food preparation, their duties in terms of paid and unpaid labor, and even daily exercise routines and entertainment: "Coolies must be encouraged to amuse themselves by harmless diversions, and should be allowed to play on their drums &c., till eight bells."[49] Needless to say, and as scholars have shown, extensive rules were enacted to manage and monitor the life and labor of indentured Indians on the plantations.[50] While the plantation, as the locus of labor extraction, seems an obvious site of regulation, such was not the case with the embarkation depot. It was largely the new bureaucratic rationality, combined with more contingent logics, which produced this particular outcome. Indeed, over the years, sub-depots were also set up along the route from the interior recruiting districts to the coast but, while subject to some regulation, they were not subject to the kind of minute scrutiny accorded to the coastal depots. As a result, we know less about conditions that prevailed at these sites.

The spatial focus of the bureaucratic gaze and the organizational transformations that came into existence were reflected in the manner in which the rules themselves came to be written up and organized. In this regard, there was a distinct change by the 1870s, as can be evinced in the set of Rules that came into effect with Act VII of 1871.[51] Whereas the Rules in 1864 were outlined as lists of duties of the different officers in charge—namely, the Protector of Emigrants, the Medical Inspector, and the Emigration Agent—in 1871 they were organized temporally.[52] Now the process of migration was temporally demarcated and specific spaces attached to different sequences of the process. The four sections included in the contents of the rules for 1871 were as follows: "Of Preliminary Arrangements in Port," "Of the Embarkation," "Of the Voyage," and "Of Returned Emigrants."[53] By 1883, a new section on "Recruitment" had been added and, by 1892, the process of migration had expanded to include seven distinct stages: "Chapter I: General Rules," "Chapter II: Recruitment," "Chapter III: Depots," "Chapter IV: Transport Arrangements," "Chapter V: Voyage," "Chapter VI: Returned Emigrants," "Chapter VII: Records," and the "Appendix."[54] The duties of the several officers were now outlined under these particular time–space articulations. This particular spatio–temporal, linear organization of migration, which derives from a bureaucratic and legalistic logic, is now also one of the predominant frameworks for studying migration. The regulatory structure has been mapped onto the knowledge structure, or the categories of practice have become indistinguishable from the categories of analysis.[55]

Health, Disease, and the Migrant Body

The overarching rationale that anchored the proliferating rules in the regime of migration control was the "wellbeing" of the emigrant. In the lexicon of this regime, "wellbeing" amounted to a narrowly circumscribed notion of simply being alive.[56] There was a preoccupation with the body, largely motivated by a barely disguised interest in its ability to serve as an efficient source of labor, and the prerequisite for a nominally healthy and living body. The body of the emigrant largely occasioned interest in terms of totalities and to the extent that it strengthened or weakened the totality, or the abstractions, of the *categories* of health, disease, and mortality. In the vast official records, there is something of a grave disregard for the *experience* of health, disease, or mortality of the individual, or empirical, body as, for instance, in the paucity of descriptions that might betray an attention to an actual body in discomfort or in pain—or, for that matter, in comfort and good cheer. Simultaneously, there is an obsessive concern with the functioning of the body *qua* body. The rules that were established besieged the body in intimate detail, permitting the constant monitoring of its fevers, its bowel movements, its precise placement in ships and depots, its ability to infect the healthy, its clothing, its diet, its exercise regimens, its hygiene:

115. The coolies must be kept as much as possible in the open air,—they should take their meals on deck whenever this is practicable, and at least one-half of their number should *always be on deck* when the weather permits it.

116. Ventilation, and the means for promoting or modifying it, should be constantly attended to; and those openings which it would be dangerous to keep open promptly closed on the approach of bad weather.

117. When possible, the coolies should bathe daily, a sail-screen being put up for the women; and they should wash their clothes at least twice a week. The decks should be *dry holystoned* by the male emigrants daily, when practicable, and disinfecting powder must be freely used where the decks have been soiled in any way.

118. The water-closets must be constantly inspected throughout the voyage.

119. The coolies' clothing and blankets should be aired thoroughly at least thrice a week. In damp weather the hanging stoves and hot sands may be cautiously used with advantage to hasten the drying of any parts of the deck, &c., that have been accidentally wetted.

120. At the outset of the voyage the Surgeon should take an account of, and hold in reserve, all the warm clothing in the possession of the emigrants, in order that they may be preserved and used as soon as a change in temperature renders it necessary. Each emigrant should be made responsible for the care of his or her own suit. There should be a general muster of all clothing and blankets once a week.

121. The provisions issued daily must be divided into two portions, so that they may be cooked for two meals; the first at 9 A.M., and the other at 3 P.M. The issues shall be recorded daily by the officer in charge, and attested by the initials of the Surgeons and officer. (See form No. 22 or No. 23). . . .

126. Coolies must not be allowed to retain or secrete in their clothes, &c., any food which they were unable to finish; the eating of stale food must be prevented. No smoking shall be allowed between decks.

127. Food for the sick, their medicines, &c., should be methodically arranged, so that they shall receive the prescribed quantities regularly every four hours, or at such intervals as the Surgeon may decide on. *The night nursing and feeding of the sick should be most practically provided for.* It is between midnight and morning that the want of sufficient support is so severely felt, and is so likely to be neglected by a careless assistant. . . .

129. . . . It is the Surgeon's duty to keep a watchful eye over the health of the coolies; he should muster and inspect them twice daily, and by being constantly among them, he should be always in a position to detect the earliest symptoms of disease, such as scurvy, &c. Communicable diseases (smallpox, cholera, &c.) must be isolated *at once*; the place vacated must be thoroughly

cleaned and disinfected; and any suspected or infected articles should be completely destroyed or thrown overboard.[57]

This is but a small sampling of the literally hundreds of numbered rules established. An ever-intensifying obsession with the health, disease, and mortality of the emigrant's body would come to rival, indeed even partially eclipse, the concern with consent and the emigrant's will. While the latter was efficiently managed and contained through the technology of the contract, the technologies for managing the body were infinitely more variegated and complex. Concerns related to the body were key determinants in establishing the logics of examination and surveillance, of the appointment of different functionaries and officers in the grid of control, and of the spatio–temporal reorganization of migration. Enmeshed in the crevices of the system, these techniques permitted an "uninterrupted, constant coercion, supervising the processes of the activity rather than its result . . . exercised according to a codification that partitions as closely as possible time, space, movement."[58]

Close attention to health, disease, and the body percolated through every aspect of the migration regime and saturated all dimensions of the regulatory apparatus. It was cause for the appointment of a variety of medical officers who were charged with a range of tasks. Perhaps one of the more important was the task assigned to the Medical Inspector at the embarkation depots: in addition to monitoring the health of the emigrants as well as overall hygiene and sanitation at the depots, he was responsible for providing an individual assessment of the medical fitness of each emigrant for undertaking the voyage to the destination colony. The Inspector was categorically *not* responsible for making an assessment as to the health and fitness of the emigrant to perform labor; the responsibility for this assessment was assigned to the Emigration Agent employed by each colony.[59] The medical certificate, testifying to the emigrant's fitness for the voyage, became, in addition to the labor contract, a mandatory requirement for every emigrant under indenture.

Yet another task that fell to the Medical Inspector was to oversee the arrangements on the ships. The Medical Surgeon in charge of the ship had to submit a report of the voyage to the Medical Inspector at the end of the journey. Surgeon Partridge, in 1863, found the Returns he received for the preceding year to be "of little utility" and "carelessly prepared."[60] To "rectify" this state of affairs, Partridge proposed precise forms he wished the ship's officers to complete. In his view, their time was better spent attending to the needs of the emigrants than writing detailed reports:

> I have drawn up a form of Return which I should like generally adopted. I think it very important that the Surgeon Superintendent of an Emigrant Ship should have as little to do in the way of writing,—his whole time should be given up to the personal supervision of the individuals entrusted to his care, and it is, I believe, in exact proportion to his activity in this respect that the voyages he makes are successful or the reverse; entertaining these views I have *tabulated*, as far as possible, the items concerning which I require information, and I am convinced a few minutes' attention daily will suffice to collect the data necessary to fill up the Return when the voyage comes to a close.[61]

The templates he offered simply required minimal information of births and deaths (with a list of diseases listed as causes); a form listing the numbers of emigrants who arrived at the destination in "health," "convalescence," or were "sent to the hospital"; a weekly log of "Admissions to the Sick List and of the Deaths during the Voyage"; and two templates that required the Medical Officer to keep a daily log of the temperature "in the between-decks" and the direction of the wind and the dates of arrival at key points in the journey (e.g., the Equator, the Cape of Good Hope).[62] Partridge's suggestion of prescribed forms was adopted and, over time, supplemented. The forms he proposed and the rationale he deployed reflect the deeply entangled nature of the concern with health, mortality, and the body with the modalities of supervision, the hierarchical grid of officials that formed the structure, and the abundance of records and forms that permeated it.

Alongside the innumerable rules relating to the tasks and activities of the different officials and of the emigrants were innumerable rules relating to the minutiae of such diverse subjects as dietary provisions, medical supplies, the rate at which the distilling apparatus functioned, and the standardization of measurement techniques. Taken together, these practices— and the logics they simultaneously reflected and helped consolidate—are crucial elements of the disciplinary apparatus that produced the modern body.[63] The hundreds of rules that formed part of the migration regime came into effect either due to the keen observations and suggestions of particularly zealous officers or after in-depth consideration by committees specially appointed for the purpose and in consultation with a grid of officials and offices, who spanned countries, continents, and empires. The logic of disciplinary power thus not only was palpable and operational across the globe, but literally took shape through *direct* conversations and consultations between functionaries spread across a range of locales.

This is evident in the ways in which some of the rules regarding dietary scales took shape. Every item that appeared on the dietary scale was thoroughly examined from every angle, ranging from its nutritional value to its capacity to cause annoyance.[64] Asked for his views on revising the dietary scales for voyages to the West Indies, one Medical Superintendent, Dr. Payne, was unequivocal in recommending the elimination of tamarind and betel nut from the provisions. The former since, in addition to having "no antiscorbutic virtues not abundantly found in other components of the diet," was "calculated to act mischievously" due to its "laxative properties"; the latter because it was "of no value, and its use, by leading to the staining of the decks [due to spitting], would bring the Emigrants into conflict with the crew of the Ship; a result to be most carefully avoided."[65] Mortality, of course, was especially closely scrutinized and attempts made to decrease the often abysmal trends. In analyzing the high mortality on board the *Wellesley* to Trinidad in 1857, particularly of infants, 44.83 percent of whom had died, one Dr. Mouat presented a brief comparative assessment with the mortality of infants in England, Wales, and Italy that ranged from 22 percent to 36 percent.[66] He thus concluded that the mortality rate of 44.83 percent "great as it is, is not much in excess of the ordinary risks of life among infants on shore."[67] Nonetheless, he recommended a series of changes to the dietary scales to incorporate special provisions for infants and nursing mothers. As part of his rationale, Mouat once again provided a comparative assessment with regulations in force in the case of European mothers and children, included the expert opinion of "one of the most eminent obstetric practitioners in London," and outlined an extensive justification for the changes:

14. When the secretion of milk in the mother is arrested by violent Seasickness, or other similar causes, and this is followed by a diet unsuited to nursing, the secretion is scarcely ever restored.

15. When to this again is added the fact, that no suitable diet of any kind is provided for infants or very young children in Coolie Emigrant Vessels, it is more surprising that any survive, than that the mortality among them should range between 20 and 90 percent, as it did in the Ships of the Season 1856–57. . . .

25. In addition to the ordinary rations shipped, I recommend that for every native mother with a child at the breast, a pint of preserved milk be allowed daily over and above her ordinary scale of diet; and for every child under two years of age who has no mother, or whose mother is unable to nurse it, a daily ration of a pint of preserved milk.[68]

The dietary scales on emigrant ships thus came to incorporate special provisions for nursing mothers, infants, and the sick.

At the time that Mouat offered these recommendations for special provisions for nursing mothers and infants, he held the position of Inspector of Jails of the Lower Provinces in British India. The cross-fertilization and exchanges—of personnel, of techniques of control, and of expert knowledge—between jails, the military, and the emigration bureaucracy was a common occurrence.[69] In 1861, for instance, Dr. Payne noted the sufficiency of the dietary scales prescribed for passengers on emigrant ships since they were similar to those used in the jails.[70] In 1869, the high mortality on ships resulting from an epidemic was cause to circulate, in pamphlet form, to the Surgeon Superintendents of ships the "excellent descriptions of the disease given by Dr. W. Walker, who was in charge of the Agra Central Prison during the epidemic of 1860, and Dr. Grey, who was Superintendent of the Lahore Central Jail during the latter part of the outbreak of 1863."[71] In 1871, the president of the inter-empire Special Committee appointed to revise the rules for the provisions sent on French ships was the Head Physician of the military hospital at Pondicherry. The British Consular Agent in the French colony, who contributed to the committee, was a military official, Colonel Doveton, and all meetings of the committee were held at the military hospital.[72] The techniques of disciplinary power were both developed and deployed extensively at each of these sites; indeed, techniques developed at one site would often quickly find their way to another site, either through overt modes of transfer and exchange, such as those recounted above, or through more discreet and imperceptible mechanisms.

From our perspective, the important point is how the multiple techniques and mechanisms of disciplinary power not only proliferated through the nineteenth-century apparatus of colonial emigration control, but pervaded also the workings of the colonial jails and military, among other sites. The modern body, both target and consequence of disciplinary power, was, in this way, produced in similar fashion across these different domains. Among the significant similarities was the unceasing production of the body as an object of the distinctly modern discourses of health, disease, and mortality that were intertwined with the new logics of population, demography, and epidemiology.

Rule by Record, Record of Rule

Under the logic of surveillance and examination, the interminable process of instituting rules to manage Indian emigration worked in tandem with the interminable process of formulating technologies for meticulous record keeping. Every rule instituted produced mechanisms that sought to ascertain and document that it was followed. Beginning with the simple requirement of a nominal labor contract from each emigrant, over time there emerged an ever-increasing number of forms to be completed, compiled, submitted, and filed. Act XV of 1842, which resumed Indian migration to Mauritius, was a mere seven pages, including the schedule of orders.[73] When, some twenty years later, in 1864, Henry Maine attempted to systematize the process, in addition to the Act itself, it took nineteen pages to identify just the Rules for the Protector of Emigrants, the Medical Inspector, and the Emigration Agent.[74] Some ten years later the Rules that accompanied Act VII of 1871 had burgeoned to sixty-two pages, which included thirty types of numbered forms.[75] Another decade later, the Rules relating to Act XXI of 1883 bloated further, stretching to 103 pages with fifty-nine forms.[76] By 1892, in addition to the various rules and forms, there was, in fact, an entire chapter devoted to "Records" in the Rules Relating to Colonial Emigration.[77] This chapter distinguished between records to be kept in perpetuity (such as the register of emigrants recruited, or the Surgeon Superintendent's register of births and deaths at sea); those records to be kept for three years (such as the returns of the sickness in the depots, or the certificate of medical examination of the emigrants, or the Surgeon Superintendent's journal); and those records that should "be destroyed on the expiry of one year from the date of their receipt" (such as the "returns, registers, copies of agreements and other papers required to be furnished to the Protector of Emigrants or the Medical Inspector").[78]

This was but one aspect of the system of "intense registration and documentary accumulation" that accompanied the procedures of examination, a modality Foucault identifies as central to the operations of disciplinary power.[79] The functions and effects of this series of forms and documents were akin to those produced by the examination, which not only "places individuals in a field of surveillance [but] also situates them in a network of writing; it engages them in a whole mass of documents that capture and fix them. . . . A 'power of writing' was constituted as an essential part in the mechanisms of discipline."[80] The different modalities of serialization, identification, documentation, classification, and surveillance that are part of disciplinary power would all coalesce in the rules laid out for the guidance of Emigration Agents regarding the embarkation of emigrants:

32. In sufficient time previous to the day fixed for embarkation, the coolies should be arranged (relatives being kept together) in the numerical order in which their names will occur in the passes and the general register compiled from the passes, both of which must therefore correspond and be numbered accordingly.

33. Each emigrant must be provided with his pass, and may also be supplied with a tin ticket (to hang conveniently from the neck or arm where it may be easily seen), bearing the number, stamped in raised figures, corresponding with that affixed to his or her name in the pass and general register of emigrants about to be embarked. . . .

63. The emigrants, with their clothing, baggage, &c., should be at the wharf in sufficient time to be correctly arranged for medical inspection and muster, according to their numerical order in the general register, No. 1 being placed nearest the embarkation jetty.[81]

The similarities between these arrangements and military discipline are obvious, as is the obsession of disciplinary power with efficiency and infinitesimal detail. Here, every element and medium of writing, recording, and documentation itself needed to conform to prescribed patterns and complex grids of correspondence.

The emigrants were not alone in being subjected to the "power of writing." As we have seen, each position in the network was subject to the unrelenting system of registration, surveillance, documentation, and record keeping. Every functionary was assigned a specific set of tasks and was required to submit documentation of having completed the tasks in the manner specified. Deviations from the specified rules or a failure to fulfill the tasks were equally objectionable, frequently resulting in a review or a special inquiry. Further, as with the 1863 suggestion of Dr. Partridge that required the ships' surgeons to fill out a set of predesigned forms, both the description of the tasks and documentation corroborating their execution were increasingly rendered in the shape of forms and tables that simply required completion.

Over the years, the multiplication of lists, forms, and tables combined with other dimensions of knowledge facilitated another aspect of disciplinary writing with regard to migration control: "the correlation of these elements, the accumulation of documents, their serialization, the organization of comparative fields making it possible to classify, to form categories, to determine averages, to fix norms."[82] Part of the impetus for gathering information and organizing knowledge in the shape of standardized forms and tables was precisely to achieve consistency and enable comparisons. The various forms, tables, lists, returns, and reports made possible all manner of space- or site-specific comparisons—between colonies, between depots, between ships, between plantations—and across time for a host of variables. These, in turn, were productive of new knowledges that were generative of new sites of regulation, new modalities of control, and new categories of observation that necessitated new norms that took shape as ever more detailed sets of rules and records. In other words, once put in motion it seemed impossible to constrain or contain the perpetual expansion of rules, and the records they required, that formed the disciplinary apparatus for managing Indian emigration.

The Rule of Rules: Disciplinary Power

Given the massive bureaucracy and the minute rules that monitored Indian migration, how might we explain a curious triad of facts: first, that the system was acutely violent, riven with abuse and exploitation; second, reports of such abuse and exploitation, rather than being infrequent, were a regular feature of the system; and, finally, despite the constant reports of abuse, the system continued, largely undeterred and unrevised, for almost a century.[83] One explanation is that the system and thus the rules themselves were inherently exploitative, even if this was not evident to contemporaries. Closely allied is the position that the system survived due to the economic imperatives of plantation economies and imperial necessities. Both explanations are important.[84] Another explanation, constantly offered by the contemporary bureaucrats of migration themselves, was that the abuse was an aberration, the result of the rules not being followed, or that more extensive rules were needed.[85] Within the domain of disciplinary power, the greatest infraction "is nonobservance [of the rule], that which does not measure up to the rule, that [which] departs from it."[86] Discipline thus solicits and demands conformity to the rule, though it might not operate to normalize.[87] The questions it incessantly poses are: What are the rules? Are the rules followed? Are the rules sufficient?

In close to a century of Indian indentured migration these questions recurred in the daily, humdrum operations and communications of the bureaucracy and were also the questions repeatedly raised in the face of extraordinary calamity. In the register of the mundane, for example, was what one might call the application of the rule in reverse, with the Emigration Agent at Madras informing the authorities at Mauritius that, *despite* its benefits to emigrants, ships need not have sleeping berths or platforms raised six inches off the deck since it was not "clearly expressed" and thus not mandated by any rule of Act XV of 1842.[88] In the register of the extraordinary calamity, for example, was the death of 262 persons, more than half the passengers setting sail on their voyage to British Guiana in 1865. The "searching inquiry" into the disaster concluded that the loss of life was attributable to the fact that many of the crew seemed to be sick and the Protector of Emigrants had not followed the rules. Additional rules were necessary: "The practical upshot of this lamentable case," wrote Mr. Geoghegan, "was that the departure of emigrant ships from the Mutlah was forbidden, the employment of none but first-class tugs stringently enjoined, and rules passed distinctly binding both Protector and agent to muster the crew and ascertain their efficiency, a medical certificate as to the health of the officers and crew being also required. The examination of life-buoys and boats was also enjoined by a special rule."[89] In this way, at least five new rules sprang from the "lamentable case." In addition, the Protector was "severely rebuked" for his negligence; meanwhile, 262 people were dead. It is ironic that each of the two cases described above occurred on the heels of the enactment of major emigration legislation (Act XV of 1842 and Act XIII of 1864, respectively) and thus directly followed a well-documented extensive assessment and augmentation of rules already in place. Each of these examples, moreover, can be endlessly multiplied.

If the justification for the institution of rules was the "wellbeing" of the emigrant, the implicit logic of the system was that a proper set of rules could, on their own, in fact produce the outcome of "wellbeing" and ensure that exploitation and abuse would diminish, if not disappear. "Wellbeing" was thus a procedural matter that required following rules and, more importantly, producing a record that they were followed.[90] However, "wellbeing" was not the consequence of the ever-increasing set of rules. In fact, though the infraction of the rules produced inquiries or rebukes of the officials concerned, they did not produce anything more dramatic than more rules.[91]

In time, the rules themselves came to incorporate also rules for exceptions. This is in the nature of *every* modern bureaucracy, an aspect of disciplinary power that has not received sufficient attention. One might say that every disciplinary apparatus has a series of internal limits that allow for the exception.[92] Even as these are provisional limits, called exemptions, which operate by way of temporary suspensions, specific penalties, extenuating circumstances, convincing reports, bureaucratic discretion, they also allow for the systematization and routinization of the exception itself, as was the case with the rules on Indian emigration. Hence, alongside the prescription of dietary scales was a rule permitting exceptions that took into account the state of the market and the availability of goods. Alongside the rule that native surgeons could serve only in the capacity of assistants was a rule that outlined the circumstances under which an exception could be sought. Alongside the rule establishing the proportion of male to female emigrants to be recruited for each voyage was a rule that established the conditions under which this requirement could be temporarily suspended. A specific form was devised for the use of Emigration Agents applying for an exemption from this rule. The existence of a prescribed form is a clear indication of the routinization, one might even say normalization, of exceptions themselves.

Conclusion: Disciplinary Power in the Colonial Field

One of the characteristics of sovereign power is its overt, even spectacular, nature. By contrast, disciplinary power is humble, almost unnoticeable, deploying a reverse logic of visibility: rather than being seen, it is a power that sees, that surveys. It is perhaps in keeping with the operation of disciplinary power that it can also take shape imperceptibly, without raising hackles or causing much debate. It certainly took hold of Indian emigration and produced a massive bureaucracy almost without comment. This is not to suggest that Indian emigration itself proceeded without debate, but, rather, that the kinds of debates rarely, if ever, targeted the *techniques*, the *forms*, or the *logics* of disciplinary power, even if they targeted and debated the specific *content* generated by these logics—such as the precise length of the work day, the specific proportion of male to female emigrants, the details of dietary scales, or the exact method for supervision of the sick. In a curious way, such discussions and debates worked to strengthen the force and extend the reach of disciplinary power, since the consequence of the modalities of review and inquiry—which almost always resulted from any disagreement, debate, or disaster—were further rules and regulations. Indeed, I have argued elsewhere that, in relation to Indian indentured migration, the modality of the "inquiry"—a version of the examination—served to facilitate rather than hinder the enterprise, that indenture continued not *in spite* of the endless inquiries but *because* of them.[93] The apparatus of Indian emigration control was enmeshed in and deployed a range of techniques usefully characterized as disciplinary power, which, as I have suggested, did not displace or replace sovereign power but infiltrated it. In the regime that managed Indian indentured emigration, the onslaught of disciplinary techniques and mechanisms did not cause modalities of sovereign power to disappear or even recede. Indeed, if the contract and corporal violence are two of the most explicit expressions of sovereign power, then, in this instance, disciplinary power and sovereign power served to augment rather than dislodge each other.

This chapter sought to analyze the making of a state-controlled migration regime and extensive bureaucracy that emerged through complex transactions across the globe and not as the result of simply "importing" or "imposing" a ready-made template originating in Europe. Instead, the extensive bureaucratization and minute regulation of Indian indentured migration were organic processes, simultaneously necessitated by context-specific exigencies even as they deployed modalities of regulation that had previously animated both colonial and metropolitan schemas in other arenas (such as the military or the prison). The regulation of indentured Indian migration offers an instance of a thoroughly modern bureaucratic formation that made extensive use of disciplinary power in diverse parts of the colonial world and thus invites us to consider the colonial genealogy of the modern state.

However, despite the intense micromanagement of indentured migration, important patterns of mobility, both large and small, remained outside the jurisdiction of sovereign authority. In the period under consideration thus far, the bulk of Indian labor migration took place outside state control,[94] as did the smaller migrations of merchant communities.[95] In short, the state did not exercise a monopoly over migration practices, with the regulation of indentured migration, or migration under contract, retaining its exceptional nature. As we shall see in the next two chapters, in the early twentieth century patterns of movement unregulated by the state would generate deep racial anxieties in white-settler colonies. These anxieties would be resolved, in piecemeal fashion, by new articulations of sovereignty. Specifically, such new, ad hoc resolutions would produce a notion of *national* sovereignty and catalyze the emergence of a state monopoly over migration. In the process they would remake the state and the interstate system itself, by *generalizing* what had emerged, with the institution of the contract, as an *exceptional* articulation of state sovereignty.

CHAPTER 3

Gendered Nationalism, the Racialized State, and the Making of Migration Law

The Indian "Marriage Question" in South Africa

Viewed from the vantage point of our contemporary moment, there are at least three striking features of nineteenth-century regulation of indentured Indian migration. Two of these features I have explored in the previous chapters: the logic of facilitation that governed migration and the absence of a state monopoly over migration. These features are in stark contrast to current regimes that are overwhelmingly governed by logics of constraint and where movement between countries that is not under the direct supervision and purview of the state is, ipso facto, illegal. The third and perhaps most striking feature of regulation of Indian migration is the near-absence of a discourse of nationhood, nationalism, or nationness framing discussion and action before the late nineteenth century. This absence is especially striking since concerns coalescing around gender, sexuality, and sexual morality, articulated precisely to anti-colonial Indian *nationalism*, became a critical lever in accomplishing the end of indentured migration.[1]

This and the following chapter explore the place of migration in how a world dominated by empire-states was transformed into a world dominated by nation-states. A range of historical events and processes would (re)make and produce notions of nationhood and the modern nation-state through complex circuits of connections that not only moved between metropole and a given colony, but also, as with practices relating to indentured migration, placed different colonies in relation to one another.[2] Foregrounding such circulations and relations, an organizing feature of imperial formations, these chapters turn attention to the formation of a tight confluence between nation, race, and state that would produce migration law and regulation within logics of restriction. Race-based thinking, legislation, and policies were not new; what was new was the specific articulation and institutional forms generated by the convergence of race and nationness. The particular form of indentured Indian migration can be attributed to an economic imperative, wherein a racial or civilizational rationale worked in concert with a capitalist rationale to meet the labor demands of a plantocracy. The analysis of non-indentured or "more free, 'free' migration" presented in these chapters demonstrates the relative autonomy of racial thinking that would be mobilized even when in conflict with, or unrelated to, a formal capitalist logic. Moreover, though racial thinking retained a relative autonomy from a capitalist logic, it was intimately braided with and took shape through tropes of gender and sexuality.

In this chapter I trace, in particular, how gendered discourses, sexual arrangements, and the reorganization of kinship were entangled in the race-based migration regulations attendant on the transformation of the empire-state into various nation-states. Such relationships, as I elaborated in the introduction, have been neglected by influential "developmentalist" accounts that are ill suited to analyzing the workings of gender in the formation of nationness, nationalism, and the nation-state. The scattered, uneven spatio-temporal logics of the nationalization of migration, particularly the dynamics of gender, are better apprehended by "eventful" approaches more attuned to grasping the often unpredictable, transformative trajectories of historical occurrences.[3] The specific event I consider here is the debate that erupted, between 1911 and 1914, over the legal recognition of polygamous Indian (i.e., Muslim and Hindu) marriage in South Africa. I show how, after a brief spell within the field of an imperial liberal-juridical discourse, public debate was completely reinscribed within the field of a gendered discourse of Indian nationalism. I argue that this reinscription was both provoked by and fundamentally redefined the last phase of Gandhian *satyagraha* (or passive resistance struggle) in South Africa and, simultaneously, shaped multiple, interrelated trajectories: it helped secure the South African state's race-based migration policies within the new logic of nationality, aided the formation of migration law construing monogamous Christian marriage as definitive of legal marriage, and produced the state-authorized marriage certificate as one of the regulatory technologies of migration control.

The chapter proceeds in four sections. The first offers a schematic outline of the relevant political–legal contours of Indian migration and settlement in southern Africa from the nineteenth century, including the emergence of the satyagraha struggle, under the leadership of Mohandas Karamchand Gandhi. The second section provides a genealogy of court cases regarding the "marriage question" as they related to significant aspects of migration (as distinct from settlement) regulation. The next section details how the activation of a densely gendered Indian nationalism yoked the "marriage question" to the satyagraha campaign and outlines the resolution this articulation achieved with respect to the migration of Indian wives. Finally, before turning to the conclusion, the fourth section considers the recalibrations and convergence of state discourse—in India, South Africa, and England—that affected a religious–racial nationalization of marriage relations. Thus, through an analysis of the events surrounding the legal recognition of polygamous Indian marriage in South Africa, I show how the state deployed a mechanism of gendered access to mobility to secure its race-based migration policies, and how the eruption of a gendered discourse of Indian nationalism intertwined with the politics of satyagraha aided, rather than undermined, these goals of the state. By wrenching the regulation of marriage from the jurisdiction of religious authority into the jurisdiction of state law, these events helped redefine marriage and nationalize kinship relations.[4]

A Brief History of Indians in Southern Africa

The controversy over the Indian "marriage question" in South Africa unfolded within a heterogeneous political–legal field, constituted by multiple and dissonant strands, which characterized South African legal culture in the first decades of the twentieth century. Prior to the South African (or Anglo–Boer) war of 1899–1902, the political status of the British colonies of Natal and the Cape and the Afrikaner republics of the Transvaal (the Zuid-Afrikaansche Republiek, i.e., the South African Republic) and the Orange Free State within the larger structure of the British empire-state differed widely. After the war, the four colonies or provinces came together, if reluctantly, as the Union of South Africa, under responsible government, in 1910.[5] The different colonies had, hitherto, each pursued different migration and settlement policies. Thus, in terms of the legal distinctions between the conditions of mobility, residence, and domicile, the Indian population in South Africa from the later nineteenth century can roughly be divided into three groups: (1) indentured Indians who, beginning in 1860, were specifically recruited for labor in Natal; (2) ex-indentured Indians and their descendants, the colonial-born Indians; and (3) predominantly Muslim and Parsi Gujarati traders who had migrated and settled in the different colonies at their own expense. As a result, Indians in South Africa constituted a highly class-stratified, linguistically and religiously diverse population. Specific elements of the legal distinctions of the conditions of mobility and residence of the various Indian immigrants would come to have a direct and indirect bearing on the "marriage question" and the events of 1911–14 and, therefore, are important to rehearse.

As with Mauritius and the Caribbean earlier, Indian indentured labor first moved to Natal to work on sugar plantations. Over the years, indentured labor also worked in the coal mines in the north as well as in the construction and maintenance of the Natal Government Railway. Due to a combination of factors that included reports of gross abuses in the system, the world depression following the American Civil War, and a decline in the demand for sugar in Great Britain, the migration of indentured labor to Natal was suspended in 1866.[6] The practice was resumed eight years later in 1874 and eventually terminated in 1911. It is estimated that during the operation of the system 152,184 Indians came to Natal under indenture.[7]

In many respects, the general outline of the system, as well as the terms of individual contracts governing indentured migration to Natal, was similar to that of other colonies.[8] For instance, it included the stipulation for a certain minimum male-female ratio, set at four women for every ten men; it allowed indentured laborers to have a free passage back to India at the end of ten years' "industrial residence" in Natal; and, until 1891, ex-indentured laborers could, in theory, forfeit the cost of their return passage for the grant of Crown lands in Natal. While these were some of the formal terms regulating indenture, in practice the system worked differently. For instance, Natal was loath to have women migrants since this increased the possibility of a settler population. The intent of the low female ratio in the regulation (which was frequently not met) was precisely to prevent, or certainly to restrict, settlement.[9] Further, though 52 percent of the more than 150,000 people who moved under indenture stayed on in southern Africa, and thus never used their return passages, only a total of fifty-three grants of Crown lands were ever made.[10] The practice of granting land rights substantially upset the articulation between white supremacy and class privilege, and in an attempt to strengthen this link and placate white sentiment, the system of granting land rights to ex-indentured Indians in Natal was eliminated in 1891.[11] Four years later, in 1895, Natal imposed an annual £3 tax on every adult who had either come to the colony under indenture or was the descendent of an indentured migrant who had opted out of the indenture system. In 1903, the tax was extended to include every girl over the age of thirteen and every boy over the age of sixteen.[12] The aim of this tax was to force laborers either to reindenture, in order to avoid the tax, or to leave South Africa. The Government of India had agreed to the tax so long as the Natal government would not impose criminal penalties for nonpayment. Thus, in practice, the tax operated quite unevenly.[13] Despite its uneven operation, the tax would become, if unexpectedly, a crucial aspect of the 1913 satyagraha campaign and the events related here.[14]

Though Natal was the initial destination of indentured Indians to South Africa, many thousands had drifted into the Boer territory of the Transvaal at the end of their indenture and before the South African war of 1899–1902. Hence, in the Transvaal too there emerged a population of ex-indentured and colonial-born Indians as well as restrictive policies guided by ideologies of the "Asiatic menace." Natal was seen by the other colonies as the primary source of this menace and, particularly after the formation of the Union in 1910, was under immense pressure to terminate Indian indentured migration and thus create the minimal conditions for drafting the first Union-wide Immigration Bill.[15] Ultimately, the Government of India, seemingly unilaterally, suspended indentured migration to Natal in 1911.[16]

As noted earlier, in addition to indentured and ex-indentured Indians, South Africa had a small but politically and economically conspicuous population of Gujarati Muslim and Parsi traders who had come either directly from India or from other colonies, such as Mauritius, at their own expense.[17] Within the classification schemes that emerged in southern Africa, these traders were referred to as "free passengers," as distinguished from those who came under the state-managed Natal scheme of indentured migration. Unlike the indentured Indians or even the ex-indentured, or "free" Indians, the free passengers moved back and forth between South Africa and India. This was particularly true of the wives and children of the traders. Thus, the conditions of mobility and residence of the free passengers were markedly different from those of the indentured and ex-indentured laborers. For instance, in Natal, they were not subject to the annual £3 tax and many men qualified for the colony's sex- and class-based franchise until 1897, when franchise qualifications were altered to exclude Indians.[18] In the same year, two other pieces of legislation were introduced in Natal, both directly targeting free passenger Indians and ex-indentured Indians of the petty bourgeoisie: the inclusion of a European language requirement into the immigration regulations for free passengers seeking entry into Natal and the ratification of the Dealers' Licenses Act, which regulated the issuance of trading licenses.[19] The most objectionable part of the Dealers' Licenses Act was its reliance on what Martin Chanock identifies as "administrative law"—law dependent on administrative or bureaucratic discretion.[20] Applicants refused trading licenses under the Dealers' Licenses Act had no recourse to the courts, and the decision of the licensing authorities was final.

While Natal was the only southern African colony to employ Indian indentured labor,[21] Indian free passengers arrived at a variety of ports and many extended their enterprises into all four provinces. They soon formed a community throughout South Africa that was closely knit together by a network of business interests and family connections.[22] However, like the indentured and ex-indentured Indians, the bulk of the free passenger population was also concentrated in Natal and Transvaal. The Orange Free State had the lowest population: an estimated nine Indians had annual trading licenses in 1890.[23] The presence of these nine Indian traders in 1890 served as sufficient cause for the Orange Free State to pass legislation disallowing Indians from owning land or carrying on trade.[24] In 1911, following the formation of the Union, the Orange Free State had a mere 106 Indians.[25] After the Orange Free State, the Cape Colony had the next lowest population of Indians, an estimated 6,606.[26] Though it prided itself on its "liberal" policies and governance, in 1892 the Cape Colony had altered its relatively non-restrictive franchise law to withhold the franchise from those men unable to write in the English language and raised the accompanying property qualification as well. The British imperial state had ratified the law since it did not directly cite race as the basis for (dis)qualification.[27]

In sum, by the close of the nineteenth century, each of the four white-settler colonies in southern Africa—Natal, Transvaal, the Cape, and the

Free State—had instituted all manner of discriminatory legislation against "Asiatics." An ambiguous and malleable category, open to a series of historical redefinitions, in late nineteenth- and early twentieth-century South Africa it was primarily used to describe migrants from India and the smaller population of migrants from China. The definition of Indians as Asiatics, premised on a notion of geographical origin, existed in constant tension with their simultaneous definition, in political terms, as British subjects. In theory, this political relation posited the legal equality of British subjects throughout the world. As with the Queen's proclamation in 1858 that brought India under the direct rule of the British Crown, the first condition when the British annexed Natal in 1843 stated that "there shall not be in the eyes of the law any distinction of colour, origin, race or creed; but that the protection of the law, in letter and substance, shall be extended impartially to all alike."[28] Thus while discriminatory legislation against Asiatics in the different southern African colonies was driven by a racialized logic, it was frequently disguised, especially when pertaining to Indians, via mechanisms such as a language requirement or on grounds of public health, due to the liberal imperative of the British empire-state to disallow overt race-based discrimination among its subjects.

While all four colonies pursued policies against Asiatics, the laws enacted in the Transvaal against its larger Asiatic population became the pivot of political organization and agitation in the early twentieth century.[29] In 1885, the Transvaal had passed Law 3, legislation that denied Asiatics (thus including Indians) rights of citizenship, confined them to certain "locations," and restricted their right to own fixed property to the "locations."[30] At the time, the Transvaal was technically under the suzerainty of Britain.[31] The imperial state had disallowed the restriction of Indians (as British subjects) to certain locations on the grounds of race but agreed to the legislation if it was framed in terms of safeguarding public health on account of the "dirty habits" of the Indians.[32] For our purposes here it is unnecessary to detail the negotiations surrounding Law 3 of 1885, or the test cases British Indians sponsored in court to challenge the law, or how the British government, though doing little to support the Asiatic/Indian position, would claim that the situation of Indians in the Transvaal was "one of the subjects of disagreement which led to the outbreak of [the South African] war," except to note that while having passed the law, the Transvaal government did little to actually implement it until 1898–99.[33]

At this stage, its attempts to implement the law were overrun by the war itself.

After the war and British victory, however, the policies against Indians contained in Law 3 of 1885 were exacerbated rather than ameliorated. A series of amendments to the Law, as well as new laws and ordinances, were adopted that successively worsened the situation of Asiatics in the Transvaal and, indeed, in the other provinces. An ordinance proposed in the Transvaal in 1906 instigated the satyagraha movement.[34] The proposed ordinance, which came to be known as the "Black Act," required all Asiatics over the age of eight years to have a registration certificate verifying legal residence in the province and, at the time of application, providing thumb- and fingerprints.[35] It required Asiatics to have the certificate with them at all times with harsh penalties, including imprisonment, for non-compliance. In addition, the ordinance resuscitated the restriction of Asiatics to the "locations" of Law 3 of 1885 and forbade all further Asiatic immigration into the province. The Indian opposition to the ordinance objected, especially, to requiring certificates and fingerprinting of women.[36] In response to an opposition taking the form of organized passive resistance, or satyagraha, the government excluded women from its purview, requiring, instead, that the names of women appear on the registration certificates of their husbands or fathers. It refused, however, to relent on the issue of registration for men. Royal assent for the ordinance was withheld in early 1906. The Transvaal gained responsible government (pending the formation of the Union) on December 6, 1906, and, no longer needing Royal assent, the ordinance was passed into law in March 1907. With this, the Asiatics—i.e., Indians and Chinese—actively began the satyagraha campaign with an Indian lawyer, Mohandas Karamchand Gandhi (later Mahatma Gandhi), at the forefront of the struggle.[37] Satyagraha in South Africa would continue intermittently until 1913. Several features of this variegated politico–legal domain, including, on the one hand, the £3 tax in Natal and, on the other, the matter of registration certificates in the Transvaal, would come to shape the 1913 satyagraha, as well as the making of and the resolution to the Indian "marriage question" in postwar and post-Union South Africa.

"What Is the Meaning of the Word *Wife*?"

In 1910, the newly formed Union of South Africa maintained strict provincial boundaries between the four erstwhile colonies as it attempted to formulate Union-wide immigration legislation and policy.[38] The so-called Indian Question would constitute one of the most persistent, troublesome, and significant issues in framing this legislation. Mahmood Mamdani elaborates how, within the framework of indirect rule that organized legal regimes in Africa, people were distinguished as belonging to either a "tribe," if they were "natives," or a "race," if they were deemed "non-natives."[39] Those deemed to belong to a tribe, in keeping with the logic of indirect rule, or what Karuna Mantena identifies as the culturalist form of imperial liberalism, were governed via their so-called customary law.[40] Those deemed to belong to a race, on the other hand, were governed by a common, yet hierarchically organized, civil law.[41] Indians, as members of a "race" and conceived as "non-native," were thus governed by ordinary civil law. As Martin Chanock observes, Indians in southern Africa "could not, like Africans, be relegated to a different legal regime, but had to be discriminated against within and by the ordinary [i.e., European] law."[42] They thus "posed many of the most difficult problems to South Africa's lawyers" and discussion on the immigration legislation opened up acute questions regarding the legal definitions of residence, domicile, citizenship, and marriage.[43]

During the period in which the contours of Union-wide immigration legislation were being debated and devised, South Africa was not treated as one immigration area; rather, the four provinces, or erstwhile colonies, continued to implement their separate immigration regulations.[44] Within this structure, the Transvaal Immigration Restriction Act of 1907 had a series of categories of immigrants described as "prohibited immigrants." Among those deemed admissible, or not prohibited, were the legal wives and children of "lawfully resident and duly registered" men. In 1911, a wife of Adam Ismail, a Muslim Indian resident of the Transvaal, attempted to gain entry into South Africa. The wife, Bai Rasul, was denied admission since inscribed on Adam Ismail's registration certificate (required by the 1907 Transvaal Black Act that had instigated the satyagraha movement) was the name of his second wife from whom he was divorced. The case eventually went to the Transvaal division of the Supreme Court of South Africa and was decided by Justice Wessels. Wessels, though of the view that marriage under the Transvaal law meant a Christian marriage, had "thought it wiser" to give "the word 'wife' . . . [a] more liberal interpretation."[45] His convoluted June 1911 verdict on the case was reported thus in the *Transvaal Leader*:

> [According to the verdict a] Mohammedan is entitled to bring in one wife only. It was not laid down that it was the first wife, because we don't draw a distinction between the first, second, and the third wives of a Mohammedan: they are not wives in the Christian sense; the fact that he was married once does not make the others less wives. But if he chooses to bring in one wife, he cannot bring in another wife as long as that wife is his wife.[46]

In early 1912, there arose a similar case when Fatima Jussat, the first wife of Ebrahim Mahomed Jussat, attempted to migrate to the Transvaal.[47] Ebrahim Jussat's second wife, Rasool, had deserted him and gone to Natal. Fatima Jussat was denied entry into South Africa since Ebrahim Jussat claimed that, despite desertion, Rasool was still his wife.[48] Using the case of Bai Rasul as his precedent, in February 1912 the same Justice Wessels dismissed Fatima Jussat's appeal on the grounds that Ebrahim Jussat could not have two wives in South Africa.[49]

These decisions by Justice Wessels drew widespread criticism. For instance, the London All-India Moslem League writing to the Colonial Office said that Wessels's decision in the case of Bai Rasul "betray[ed] a want of acquaintance with the Mohamedan law of marriage and the obligations that it entails . . . [and thus] should be amended at the earliest possible opportunity, so as to make it consistent with the religious law of Islam . . . [and] unless this is done the Mussulman subjects of His Majesty will regard the omission . . . as a direct interference with the enjoyment of their laws and customs guaranteed to them by the British government."[50] Gandhi echoed the position of the London All-India Moslem League, writing in the *Indian Opinion* (a weekly publication he founded): "Hitherto those who have more than one wife have been allowed to bring them in without any let or hindrance. If the Judge's dictum is sound law, all we can say is that it will have to be altered. In British Dominions, wherein all religions are respected, it is not possible to have laws insulting to any recognized religion flourishing under it."[51]

The South Africa British Indian Committee, also writing to the Colonial Office regarding Wessels's judgments, pointed out, first, that the Cape Malays, migrants who had lived in South Africa for generations, followed the Islamic law of marriage and thus the practice of polygamy existed within South Africa.[52] Second, the committee noted that under the registration laws of the Transvaal, enacted in 1907 (i.e., the Black Act), a number of Indian Muslims "possessed . . . registration certificates upon which are officially endorsed the names of plural wives married to them according to the rites of Islam."[53] In light of this, the committee was "not prepared to accept the dictum of the Transvaal Judge . . . that polygamous marriages . . . are of such a character as to involve moral condemnation or material disability."[54] And, moreover, the committee felt that "this action has now been taken not so much from the point of view of the ethics of polygamy as in pursuance of the scarcely veiled policy of the Union administration to place the utmost difficulties in the way of the re-union of divided families, in order to promote the exodus [of Indians] from South Africa."[55]

Three aspects of these initial objections to Wessels's decisions are worth summarizing here. First, as the last statement from the South Africa British Indian Committee clearly indicates, Indians were well aware that the practice of polygamy was being used as a pretext for causing the departure of the Indian elite. State measures directed toward the marriage and sexual arrangements of the indentured population had taken a different trajectory.[56] The elite, however, played a significant role in the upward class mobility of ex-indentured Indians by providing, for instance, access to credit or employment opportunities. With the state taking increasingly stringent measures to deny such opportunities to "free" Indians, the position and role of the free passenger elite unsettled the bond between racial superiority and class privilege in several ways. The objections of the South Africa British Indian Committee to Wessels's decisions highlighted how controlling the movement of multiple wives was the vehicle intended to achieve the exodus of Indians for racist ends.

Second, and this again is evident from the communication of the South Africa British Indian Committee, the state's capitulation to Indian male outrage over the 1907 Transvaal Black Act, regarding the requirement of registration certificates from women, had inadvertently resulted in the state acknowledging and thereby sanctioning polygamy, for many Indian men had state-authorized registration certificates with the names of multiple wives.

Finally, adding to the strength of these objections, the claim advanced most vociferously by both the London All-India Moslem League and Gandhi rested on the founding rationale of indirect rule, what the League called the "guarantee" provided by the British Empire: a policy of non-interference by the state in the "laws and customs" of the multi-religious, culturally diverse subjects of the British Empire.[57] Whereas in the numerous other instances of anti-Asiatic legislation—relating, for instance, to such measures as a language requirement for immigrants, the issuance of trading licenses, curtailments on property ownership, and stipulations for registration and residence certificates—the Indian agitation appealed to the legal equality and identity of British subjects, or the universalist form of imperial liberalism, here the objections appealed to the guarantee of the British empire to respect difference. But what we shall see, as the story of the legal recognition of polygamous marriage unfolds, is how this claim would gradually weaken, enabling the state to deem polygamy *legally*, though *not customarily*, intolerable, even as the Indian agitation would take on an explicitly nationalist tenor, crystallizing around the theme of *national* honor.[58] Their conjunction would result in a reconfiguration of the empire-state, make available the configuration of the national state, and embed, in institutional form, the logic of colonial rule in new registers.

The legal implications of Wessels's decisions were worsened a few months later by the case of *Rex v. Sukina* in May 1912. The outlines of this case were similar to those of Bai Rasul and Fatima Jussat—Sukina was married to Hassin Mahomed, a "lawful resident" of the Transvaal.[59] Hassin Mahomed had also been married to Issa, from whom he claimed to be divorced. However, unlike the earlier cases, a crucial distinction here was that Issa was not in South Africa.[60] Despite this, Justice H. H. Jordan denied Sukina entry into the Transvaal and ruled: "With us marriage is the union of one man with one woman to the exclusion, while it lasts, of all others; and no union would be regarded as a marriage in this country . . . if it was allowable for the parties to legally marry a second time during its existence. . . . [It] is clear that his [Hassin Mahomed's] marriage to the accused [Sukina] is a polygamous one, and, therefore, . . . cannot be recognized as a marriage in this Province."[61] Justice Wessels had disallowed Bai Rasul and Fatima Jussat entry into the Transvaal on the grounds that their husbands could not have more than one wife in South Africa. Jordan's decision, on the other hand, went well beyond Wessels's "liberal interpretation" of the word *wife*, and refused entry to *even one* wife of a "lawfully resident and duly registered" man into the Transvaal. By focusing on the definition of the term *marriage*, Jordan's judgment, in effect, denied legal recognition to *all* Hindu and Muslim marriages, even when monogamous in practice, since the religions *permitted* polygamous marriage. Therefore, the legal entailment of Jordan's judgment was to nullify all Hindu and Muslim marriages.

Though the matter had first drawn attention due to the judgments of Wessels and Jordan in the Transvaal, a Cape Province judgment of March 14, 1913 would extend the illegality of non-Christian marriages beyond the Transvaal and would serve as the locus of more widespread protest. Whereas in the earlier cases, Indian men had, at one point or another, been married to plural wives, this was not true for the case of *Esop v. the Minister of the Interior*, which came before Justice Malcolm Searle.[62] Here Justice Searle denied Bai Mariam, the only wife of Hassan Esop, entry into the Cape, writing in his ruling:

> What is the meaning of the word "wife"? . . . Does it mean a wife by a marriage recognized as legal by the laws of this country, or must the term be extended to embrace a so-called wife by a custom which recognizes polygamy? The courts of this country have always set their faces against recognition of the so-called Mahommedan marriages as legal unions. . . . This is not a case where merely ceremonial forms of the marriage celebration in the foreign country . . . are different from those required in this country; . . . but this is a case where the very elements and essentials of a legal union of marriage are . . . wanting.[63]

In its bluntest form, the Searle judgment, as Jordan's judgment earlier, had overnight, as it were, changed the position of Indian wives to "concubines."[64] As the Solomon Commission would later put it, the consequence of the Searle decision was that any woman married under "a system which *recognizes* polygamy . . . [had] no legal status as a married woman, and the children [of such a marriage were] illegitimate."[65] Given the extraordinary steps taken to invalidate Indian marriages on the grounds of polygamy, one would expect that it was widely practiced. However, in 1914, with a total "free" Indian population of over eighty thousand, there were *forty* cases of polygamous marriages.[66] From this insignificant number emerged the controversy I have charted that resulted in invalidating all non-Christian Indian marriages.

At the time of the Searle decision, the Union-wide immigration bill was coming up for its third reading.[67] One of the stumbling blocks to passage of the bill had been the "Indian Question" and threats of reopening satyagraha, which, at this time, had slim support.[68] The Searle decision served as the occasion for garnering support for passive resistance in the name of the honor of Indian women. In a matter of two years, from the initial 1911 Wessels decision to the 1913 Searle decision, the entire controversy over the marriage question was hijacked, primarily through the pages of the *Indian Opinion*, by an immensely successful discourse on the honor of women, which became the locus for a particular construction of Indian nationalism in South Africa. As I demonstrate below, this construction decisively gendered the trajectory of events, with far-reaching consequences for Gandhi's satyagraha movement and for migration regulations in South Africa.

From the State to the Nation: "The Honour of Indian Womanhood"

The March 22, 1913 issue of *Indian Opinion*, appearing directly after the Searle judgment of March 14, carried two editorials by Gandhi, "Hindus and Mahomedans Beware" and "Attack on Indian Religions." Both explicitly linked opposition to the Searle judgment to support for satyagraha and objected to the judgment predominantly in terms of the honor of Indian women. As the first explained: "This is not a judgment given against an individual. . . . The meaning of the judgment is that every Hindu and Mahomedan wife is in South Africa illegally. . . . This is a state of things which our self-respect forbids us from tolerating. . . . It is, indeed, a serious question for passive resisters to consider whether they ought not to include in their requirements a redress of this unthought of but intolerable grievance."[69] If the trope of the honor of Indian women is buried in this editorial within the more generalized rubric of "self-respect," the inflammatory rhetoric of the second editorial more than compensated for it:

Any nation that fails to protect the honour of its women, any individual that fails to protect the honour of his wife is considered lower in level than a brute. We know that many battles have been fought to protect the honour of women. . . . It will be nothing extraordinary if right now we sacrifice our wealth, our stocks, our businesses and start the fight. All these things are intended for our happiness. If we lose our honour, what remains of happiness?[70]

Echoing Gandhi's train of thought, the British Indian Association in the Transvaal expressed "deep distress and disappointment at the [Searle] decision . . . whereby non-Christian Indian marriages, celebrated according to the tenets of the great faiths prevailing in India . . . are invalidated and the great religions of India insulted."[71] In the Association's view, the aim of the decision was "to disturb Indian domestic relations, to break up established homes, to put husband and wife asunder, [and] to deprive lawful children of their inheritances."[72] It was therefore "of the opinion that the questions arising out of the [Searle] decision are of such vital importance . . . [that] it will become the bounden duty of the community, *for the protection of its womanhood and its honour,* to adopt passive resistance."[73] Gandhi claimed that "the Searle judgment [had shaken] the existence of Indian society to its foundation."[74] *Indian Opinion* reported that when Kasturba Gandhi, Gandhi's wife, "understood the marriage difficulty . . . [she] was incensed" and resolved to join the passive resistance struggle and go to jail, a decision in which she was joined by several "other ladies."[75] "The step," as the *Indian Opinion* stated, "was momentous."[76] In a telegram to General Smuts, Minister of the Interior (later Prime Minister), the Transvaal Indian Women's Association wrote: "The association had come to the conclusion that the honour of Indian womanhood is affected by this judgment . . . and respectfully trusts that the Government will be pleased to amend the law. . . . If the Government cannot see its way to comply with the request, they . . . will suffer imprisonment [i.e., become passive resisters and court arrest] rather than suffer the indignity to which in their opinion the Searle judgment subjects them."[77] And Gandhi, in response to the women's decision to join the satyagraha movement, commented:

The remarkable resolution of the Indian women of Johannesburg . . . marks an interesting development of the passive resistance campaign. . . .

[If] Indian women become passive resisters, they must have what is, to them at any rate, a very serious grievance. . . . We congratulate our plucky sisters who have dared to fight the Government rather than submit to the insult offered by the Searle judgment. They will cover themselves and the land of their birth, as, indeed, of their adoption, with glory, if they remain true to their resolve to the end.[78]

Objections to undertaking satyagraha for a cause supposedly affecting only women were countered thus: "Cannot men go to gaol for women's honour and their own? . . . What is needed is that men should be men enough. . . . If we take cover behind the argument that there can be no satyagraha [in a matter concerning women] and sit back with folded hands, we shall only bring ridicule upon ourselves and our womenfolk."[79] Thus, even as women were commended for pledging themselves to satyagraha, in defense of their honor, men were chided for not being "men enough."[80] To add to the gendered rhetoric and further belittle men for not being "men enough," *Indian Opinion* published an article on Emmeline Pankhurst, the prominent British suffragette, who "though a woman . . . is as manly as any man."[81]

The protests against the Searle judgment were in striking contrast, in both rationale and tenor, to the objections just two years earlier to the judgments of Wessels and Jordan. In response to these judgments, which had paved the way for the marriage question to become a flashpoint by 1913, the objections had challenged the very legitimacy of the state to interfere with religious practice; the unequivocal position then had been that "in British Dominions, wherein all religions are respected, it is not possible to have laws insulting to any recognised religion flourishing under it." But until 1913, demands relating to the marriage question had not been included in the list of grievances that warranted satyagraha. With the Searle judgment, however, the terms of protest were radically altered and coalesced around an explicitly gendered theme of the honor of Indian women; protecting and restoring this honor was, moreover, now articulated as a legitimate, indeed preeminent, cause for satyagraha. This altered position on the marriage question must not be understood as simply incorporating—as the March 22 editorial, quoted earlier, puts it—an "unthought of but intolerable grievance" into the satyagraha campaign. For, as we have seen, the grievance was hardly "unthought of"; rather, it had hitherto been thought of differently. Beyond these shifts in how the issue was framed, and buried in the impassioned gendered contours that linked the marriage question to satyagraha, the very demands of adequate redress had also been attenuated into merely asking for the legal recognition of *one* wife of a domiciled Indian male—a demand that came to known as the recognition of de facto monogamous marriages. In short, shorn of its gendered rhetoric, the demand of satyagraha on the marriage question was only to restore to the meaning of the word *wife* the so-called liberal interpretation accorded by the two Wessels decisions.

Even as the pages of *Indian Opinion* were abuzz with the implications of the Searle judgment and sought to marshal support for satyagraha in defense of the honor of Indian women, the Union-wide immigration bill was being debated in the House of Assembly, and negotiations between Gandhi and the government continued. The Searle judgment was central to these negotiations, though they yielded little. The bill was ratified as the Immigration Restriction Act of June 14, 1913. It exempted from the category of "prohibited immigrant" "the wife or child of a lawful and monogamous marriage duly celebrated according to the rites of any religious faith outside the Union."[82] But given that *all* marriages conducted under *any* religion that *recognized* polygamy had been construed, by definition, as polygamous, the Act essentially upheld Searle's court decision and deemed all non-Christian Indian marriages invalid.

Gandhi attempted to have further negotiations with the government to alter the Act—and thus not have to make good on the threat of passive resistance—with no success. Ultimately, on September 12, 1913, A. M. Cachalia, the chairman of the British Indian Association in the Transvaal, in a letter to the government, formally resumed passive resistance warning that it would expand beyond the Transvaal and include women.[83] In addition to demands relating to, for instance, inter-provincial migration, the most significant demands of passive resistance in this phase were a resolution of the marriage question and repeal of the £3 tax on ex-indentured Indians. The other major distinction in this call for passive resistance was the inclusion of women as active participants in the campaign. It is ironic that women themselves gained access to the public sphere of the political through a defense of an honor that was configured around their putative role in the private sphere.

The June 1913 Immigration Restriction Act, the legislation that replicated the Searle decision, brought but a trickle of support for satyagraha from Indian organizations, as there prevailed a general sense that it would be loosely enforced.[84] Then the October 1 judgment of the Natal Supreme Court declaring Kulsan Bibi, the only wife of Mahboob Khan, a "prohibited immigrant," since she had been married under the tenets of Islam, made it clear that the Act would be administered harshly rather than liberally.[85] The Kulsan Bibi case triggered a sudden stream of support. The South Africa British Indian Committee, for instance, saw the decision in the case as "intolerable," "likely to arouse intense resentment throughout India," and demanded "an early settlement of the grievances of the Indian community . . . especially those relating to the status of Indian wives."[86] The Anjuman Islam of Durban "congratulat[ed] the brave men and women who have pioneered the last phase of the passive resistance struggle for the maintenance of religion and the national honour of the Indian community."[87] The Port Elizabeth British Indian Association "cordially congratula[ted] the brave men and women who have already sought imprisonment for the honour of the Indian community."[88] The Indian Political Association at Kimberly and a meeting of Indians at Pietermaritzburg also endorsed passive resistance and congratulated those in prison for "defend[ing] the national honour."[89] *Indian Opinion* continued on the theme of national honor:

> In so far as this marriage question involves an insult to our religions and an attack upon our national honour, it is far more serious than that of the obnoxious tax. A nation that cannot protect its women's honour and the interests of its children does not deserve to be called by that name. Such people are not a nation but mere brutes. Even animals use their horns to protect their young ones. Will men, then, if they are men, hang back [from joining the movement], clinging to their wretched finery and their pleasures?[90]

The variety of Indian organizations endorsing the principle of passive resistance in October did so expressly on the grounds of the honor of Indian women that was seen as coterminous with the honor of the Indian nation. Indeed, protecting the honor of women was, according to *Indian Opinion*, a prerequisite for the very definition of the nation, since a "nation that cannot protect its women's honour . . . does not deserve to be called by that name."

Not only did the judgment in the Kulsan Bibi case, which replicated the

Searle judgment, galvanize a number of elite organizations to lend their support to satyagraha, but the primary aim of the movement itself, as well as of the resisters in prison, was now coded as a "defence of national honour." All the other demands, including those calling for the repeal of the "obnoxious tax," were relegated to a position of insignificance by the different organizations and *Indian Opinion* alike. Members of the South African Indian community, especially men, "if they [were] men," were mobilized to join satyagraha for "the protection of [their] womanhood and [their] honour."

The historiography of satyagraha in South Africa has pointed to the Searle decision as one of the reasons, among others, for resuming passive resistance; what these accounts fail to do, however, is inquire into how the Searle decision and the judgment in the Kulsan Bibi case served as the conduits through which national identity was congealed, the honor of India was defined, and support for satyagraha was garnered.[91] No other grievance of the South African Indian community had been cast so emphatically and successfully in terms of national honor. The barrage of racist legislation and policy had not been articulated in a discourse of national honor but had been understood primarily within the more generalized framework of the "poison of race prejudice." While a discourse of honor underwrote the practice of satyagraha—in that, to be honorable, one had to be true to one's pledge—this was expressed as a pledge with God and with oneself.[92] A resolution of the marriage question, in other words, was not merely one reason, among many, for the 1913 satyagraha. The gendered discourse of national honor that now framed the issue made it preeminent, even as it redefined satyagraha itself. Regardless of people's motivations for participating in the movement, within the ambit of organizations supporting satyagraha, such participation was now constituted as primarily concerned with protecting the honor of women and the honor of the nation.[93]

Wrenching the question away from a discursive formation that utilized the idiom of liberal tolerance and a respect for difference in the direction of a discursive field overdetermined by a gendered discourse of national honor made virtually inadmissible the articulation of positions that constituted a direct challenge to the organizing principles of the state. Thus, demands from the London All-India Moslem League and the Hamidia Islamic Society that continued to mobilize the initial idiom of a respect of difference, to press for the legal recognition of polygamy on the grounds that most subjects of the British Empire were not Christian, were completely ignored by the supporters of satyagraha and the state alike, as the notion of de facto monogamous marriage replaced this demand.[94] Moreover, within the parameters of this discursive field, the fact that Indian marriages had been targeted in order to cause the departure of the Indian elite—due to the unsettling effect their presence had on the articulation between white supremacy and class privilege—was also all but submerged. The patently racist court decisions and legislation had been coded as an attack on the honor of Indian women that was tantamount to a degradation of the national honor. A discourse of race had transmogrified into a gendered discourse of national honor, with satyagraha as its beneficiary.

This shift in the discursive parameters, however, was not managed by the singular force of South African organizations allied with satyagraha. The outrage over the marriage question not only was palpable in South Africa but had caused "intense feeling" in India as well.[95] According to the Solomon Commission, the decisions had, in fact, caused greater agitation in India since they were "seen as a slur on Indian women."[96] The potent discourse of gendered national honor activated by the Searle decision fed into anti-colonial nationalism in India, leading to public meetings in major cities that decried the situation of Indian emigrants, rallied against indenture, and objected, in particular, to the South African policy on Indian marriages. The situation accelerated the nationalist demand to put an end to indenture, with Gokhale introducing a motion in the Legislative Council to discontinue the system and demanding explanations for the fact that, in Mauritius, Indians had to follow the French laws of marriage and succession. Ratan Tata, of the eminent Indian industrial family, donated several thousand pounds to the passive resistance struggle in South Africa. Women's committees in India sent their good wishes and donated both money and crafts for the Transvaal Indian Women's Association craft fair. It was not only elite nationalists who were outraged: the trope of the honor of Indian women was at the center of a growing mass movement opposing the treatment of indentured Indian migrants.[97] In India, too, the racist motivations and implications of the decisions were forgotten, as the agitations centered on the honor of Indian women. In other words, a particular constellation of forces conspired to make national honor the defining element of the marriage question that, in turn, proved to be an especially effective strategy for mobilizing elite support for satyagraha in South Africa.

Meanwhile, some such as Lord Gladstone, Governor General of South Africa, in a confidential minute warned of "the danger which might result from a continued state of friction in South Africa . . . [which] is all the greater because of inevitable exaggerations, and the efforts of revolutionaries to use the grievances of British Indians in South Africa against the British in India."[98] It was a prescient warning: the situation of Indian emigrants in various parts of empire and the inability of the empire-state to deliver on its liberal promises would increasingly be incorporated into anti-colonial nationalist arguments for extending demands from *swaraj* (self-rule, but within empire) to *purna swaraj* (complete independence). As Jawaharlal Nehru, in his speech on becoming the president of the Indian National Congress in 1921, would explain the situation: "Our countrymen abroad must realize that the key to their problems lies in India. They rise or they fall with the rise and fall of India. . . . Surely, the only way is to put an end to our subjection, to gain independence and the power to protect our people wherever they might be."[99] The entire practice of migration was being imagined, monitored, and regulated anew—by a range of actors—within a logic where territory, nationality, and population were indistinguishable.

Let me return, briefly, to the events surrounding the 1913 satyagraha. For, despite the ferment over the marriage question and the support it generated, the struggle might not have yielded much had it not been for strikes by the indentured population. This is best understood in terms of subaltern resistance that does not fit easily into the narrative of the nation.[100] Repeal of the £3 tax on ex-indentured Indians had been included as a goal of the passive resistance movement, the first time it had made an outright demand concerning the indentured population.[101] Little attempt was made, however, to build an alliance with indentured laborers. A strike of indentured workers in the coal mines in northern Natal was announced on October 15, 1913.[102] On October 16, twelve *satyagrahis*—eleven women and one man—spoke to the indentured population at Newcastle.[103] Their visit to the coal mines at Newcastle marks the beginning of the strike by the indentured population. Within two weeks, about five thousand indentured Indian workers from the mines and railways were on strike and attempting to court arrest.[104] By November, the strike had spread to the sugar plantations in the south. If the strike in the north can be attributed to the speeches by the satyagrahis, it spread without the agitation of the satyagraha movement to the south, where literally thousands of plantation laborers refused to work.[105]

In the time the strike had been restricted to the north, the government followed a policy of non-intervention, refusing to arrest the striking workers even when they crossed the border between Natal and the Transvaal, in violation of the inter-provincial migration laws. But when it spread to the south, and it became apparent that the numbers involved could be as high as sixty thousand, the state began to crack down, first, violently, on the laborers in the south and, eventually, by ordering mass arrests of the laborers in the north. In order to house the latter, the coal mines were declared prisons and miners taken back and forced to work as part of their prison sentence. Worldwide reports of flagrant state brutality and the mass arrests rendered the passive resistance struggle a success. This success also reconfigured Gandhi as Mahatma Gandhi.[106] The Solomon Commission was appointed in South Africa to investigate the matter and, on its recommendations, the Immigration Restriction Act was amended to repeal the £3 tax and give legal recognition to one wife of a domiciled Indian male. The Indian Relief Bill of 1914 led to a short-lived suspension of the struggles between the Indian community and the South African state.[107]

From this complex series of sub-events two must be noted. First, that the strikes by the indentured population were critical, unexpected, and decisive. Second, that the October 1, 1913, decision in the Kulsan Bibi case marked a watershed moment in eliciting elite support for passive resistance. It is unproductive to conjecture about the fate of satyagraha (or, indeed, Gandhi's political career) had this case not come before the courts. At the same time, however, it is imperative that we note how the trope of the honor of Indian women and the honor of the Indian nation, given renewed impetus by this decision, served to substantially increase Indian support for passive resistance, both in terms of organizations endorsing the movement and in terms of increased numbers of people actually willing to participate as satyagrahis. Indian agitation in South Africa, aided by opposition stemming from London and India, had redefined the debate on the legal recognition of non-Christian Indian marriage from a challenge to the premises of the liberal state to an attack on religion and national honor. In the process, the racialized motivations that tied the South African state's objections to polygamy to Indian migrants and residents disappeared and marriage was rendered into the regulatory hands of the state.

Refashioning the State:
Registering Marriage, Institutionalizing Nationality

Far more than the honor of women was at play in the denial of legal recognition to Hindu and Muslim marriages. First, it meant that men had no recourse to the state in the surveillance and control of women and women's sexuality, since, in the eyes of the state, all Hindu and Muslim wives were "no different from concubines." At the same time, since women had no legal status as wives, they, too, were without means for seeking redress from the state. Instead, the state could—and did—actively prevent them (and their children) from benefiting from the material advantages of marriage, particularly around inheritance. Even as "wives" were transformed into "concubines," the legal status of men remained relatively unchanged, since it was never premised on the status of being "husbands." Further, unlike the case with indentured migration, where, if reluctantly, the state had conceded some mobility to women as individual agents, here the "free" migrant was resolutely presumed to be male and women could have access to mobility only via their relationships to men.

In adopting the common subterfuge of framing marriage as a "women's issue," the Indian opposition muted the *material* disadvantages of these court decisions for women, amplified its *symbolic* impact on the honor of women, and, if unwittingly, made common cause with the state in its disregard for the embodied lives of women. It is striking that the Indian opposition to the activities of the state in 1906–7, when satyagraha commenced with the Black Act in the Transvaal, and in 1913, as support was being mustered against the Searle decision, was framed within a conservative *symbolic* discourse of honor and insult. If the state was unconcerned with the material disabilities the legal delegitimization of marriage posed for women, the Indian opposition was largely concerned with ensuring that women would continue to be regulated by the "community," rather than suggesting that they have an unmediated access to the state. As with the 1907 Black Act, which bequeathed upon men a state-assigned identity and made women appendages to men, with the names of wives and daughters on the certificates of their husbands and fathers—a direct consequence of the Searle decision was that women's relationship to the state was mediated through their relationship to men.

Even as a range of forces converged to recalibrate the discursive elements that defined the marriage question, thus bringing it within the ambit of satyagraha, a complex range of forces was at work in making the legal recognition of de facto monogamous marriages an acceptable solution. Alongside the series of reinscriptions where religious difference was coded as racial-national difference through the vehicle of woman was the position of the empire-state—in India, South Africa, and England—which also added credibility to an organic link between nation and religion. As the controversy over the legal recognition of non-Christian Indian marriages was brewing, it came to be gradually held as a truism that it was "not reasonable to ask a Christian legislature to alter the general marriage law by legalizing marriage according to religions which recognize polygamy." This was the position of Lord Harcourt, the Secretary of State for the Colonies in England.[108] In India, Sir Syed Ali Imam, an Indian—and, significantly, Muslim—member in Viceroy Hardinge's Council, voiced a similar position, which is worth quoting at length:

> [While different] incidents of minor importance attach to the contract of marriage in different centres of Christendom . . . [there is] no manner of doubt that any marriage that has not monogamy as its basic principle can ever be held to be valid . . . in any part of Christendom. The law has its origin in the Christian faith and Ecclesiastical authority, but it affects . . . [the] validity [of] marriages contracted by non-Christians if such validity is sought in a Court in Christendom. . . . It follows, therefore, that the South African Government has considerable justification for standing by a principle that it must bow to as a Christian administration. It will be a feeble argument to advance to say that South Africa is not a Christian country. . . . To all intents and purposes it is a Christian country. . . . It is obvious then, that to ask the South African Government to give up this principle is to ask it to dissociate itself from the rest of Christendom on a point affecting in the highest degree the moral and social conception of Christian nations. This must be regarded as wholly impracticable and outside the range of a reasonable solution of a difficult problem.[109]

In South Africa, the Solomon Commission found it especially preposterous that "we should alter our marriage law" to suit "persons who have *voluntarily* come to the country [i.e., "free passenger" Indians]."[110] The conundrum now was whether to allow even *one* wife to migrate to South

Africa and how to do so without giving legal sanction to polygamy within "ordinary" civil law, which governed races, but not "native tribes." By 1913, Gandhi's demand (which did not represent all sections of the Indian community) was similar: to give legal sanction to *one* wife, though the marriage was conducted under a religion that recognized polygamy. What his position encapsulated was an unspoken acceptance of the view that it, indeed, was "not reasonable to ask a Christian legislature" to give legal recognition to polygamous marriages—even as polygamy was an important element of the "customary law" that governed "natives."[111]

If the Indian uproar over the Searle judgment had hijacked the "marriage question" onto a terrain where the missing link between nation and religion was (the restoration of) the honor of Indian women, the position of the different arms of the empire-state, in blurring the line between religion and state, effectively made Christianity the state religion in South Africa. The 1914 Indian Relief Bill, offered as the resolution to the "Indian Question," would explicitly code the state as Christian. Men were free to have multiple marriages; the state, however, would recognize only one marriage and only the children of this marriage would be deemed legitimate. Moreover, in order to be recognized, the marriage would have to be officially registered with the state. This resolution expressed a novel understanding of the principle of tolerance and the relationship between "ordinary law" and "customary law": simultaneously recognizing *and* delegitimizing the latter.[112] In this way, the regulation of marriage, certainly insofar as it related to Indian migrants and residents in South Africa, was wrested out of the control of religious authority and into the control of state authority. For the purposes of participating in legal migration on the basis of marital alliances, it became mandatory for Indians to corroborate a legal marriage as documented and verified by the state through a series of stringent regulations.[113] This was in stark contrast to the approach that had governed the marriage and sexual arrangements of the more than 150,000 indentured migrants who had arrived in Natal in the half-century preceding the formation of the Union.[114]

Conclusion

Feminist scholarship has shown that familial narratives, tropes of kinship, and dense articulations of gender are central, perhaps indispensable, to nationalist discourse; that national *identity* seems unable to express itself without resorting to idioms of gender and sexuality.[115] Simultaneously, particularly since the nineteenth century, state regulation of marriage, kinship, and filiations has become an increasingly important realm with regard to producing and policing the limits of modern notions of nationality through procedures of *identification*.[116] These twin forces—of socio–cultural formations of identity and politico–legal procedures of identification—that subtend the notion of nationality are premised upon and call forth a demand for endogamy. This mingling of family genealogy with the definition of national community, as Étienne Balibar notes, "is a crucial structural mode of production of historical racism . . . [which] is also true when the national becomes a multinational community."[117] Thus, immanent to all invocations of nationality are relations of gender, sexuality, and kinship.[118] In South Africa the implementation of anti-miscegenation laws would demand endogamy within the internally differentiated and racially classified nationalities and tribes. However, the principle also animates the notion of nationality in general. For this reason the migration of people threatens the stability of (contingent) nationalities in a fundamental way.

The 1914 South African Indian Relief Bill explicitly identified "Indians" as a national category with regard to migration regulations. Earlier the category used had been "Asiatic." In fact, as Karen Harris notes, legislation that specifically targeted and isolated Indians as a *national* group emerged only after the formation of the Union of South Africa.[119] Such transformations in the classification of people, from "Asiatic" to "Indian," from a regional category to a category understood precisely as a nationality, speak to the microscopic, almost surreptitious, global transformations of the empire-state into the nation-state. In other words, the identification (in affective and legal registers) of Indians as a national group by *both* Indians and the state fed into processes of nationalization that enabled a recoding of the racialized logic of the state as a naturalized logic of nationality. While Indians, and aspects of migration law, were "nationalized" before the emergence and consolidation of a specifically "South African" national identity, these events nonetheless invested the state with a national character by generating nationality as a viable state (and social) category.[120] "Nationality," thus, was an unforeseen outcome of these events.

As "Asiatics" were being transformed into new kinds of nationality-bearing "Indians" and "Chinese," the category of "British subject" was also undergoing a thorough redefinition. Contributing to this redefinition, by placing an enormous burden on the political relation that posited the equality of British subjects, was the movement of racialized subjects to white-settler colonies of the British Empire. As I will demonstrate in the next chapter, in tracing a history of the passport, the category of "nationality" emerged as a central feature of migration regulations precisely in order to manage the racialized anxieties that threatened to break apart the category of "British subject." This new logic of nationality served as an alibi for race and was able to preserve intact the notion of universal liberal equality.

CHAPTER 4

Race, Nationality, Mobility

A History of the Passport

In a sense, every modern nation is a product of colonization.
—ÉTIENNE BALIBAR, "The Nation Form"

At the heart of the South African state's incorporation of kinship structures into its migration regulations was a logic of restriction. Indeed, this logic was gaining ground in the early twentieth century. This chapter charts a history of a central technology that regulates migration through a logic of restriction: the modern passport, which emerges through the articulation of nation, race, and state. Given the current ubiquity of the passport as a necessary document for international mobility, one might expect that a passport system emerged, full-blown, in a world of nation-states. Its development, however, was rather more checkered, piecemeal, and counterintuitive. There are no definitive "origins" for the passport system, and, even today, it is a system that lacks systematization and standardization.[1] My focus here is the sequence of events and protracted debates between 1906 and 1917 surrounding the Canadian demand that Indians emigrating to Canada should have passports. This demand was largely made on the grounds of race, though rerouted through arguments of insufficient labor demand, cultural incompatibility, or unsuitability of the climate, and eventually accepted on the newer grounds of national sovereignty. An analysis of these events and debates demonstrates that the passport is not a technology simply *reflecting* certain understandings of race and nationality but, instead, was central to organizing and securing the modern definitions of these categories. More specifically, I am concerned here with exploring events through which nationality was historically produced and codified as a territorially delimited category and emerged as the privileged axis for state control over mobility. How did nationality come to signify a privileged relation between people and literal territory? Unlike much work on migration that contends that immigration *disrupts* fictive notions of the purity and homogeneity of the national, I suggest that migration precipitated the emergence of nationality as a staunch territorial attachment.

I shall, thus, advance two primary arguments. First, I develop further the argument already proposed in the previous chapter, that the modern economy of migration, grounded in race and imperialism, is fundamental to the creation of a geopolitical space dominated by the nation-state. This argument resonates with, even as it differs from, the work on "new racism" and the nation-state, one of whose central and most useful projects is to demonstrate, in Paul Gilroy's words, "how the limits of race have come to coincide so precisely with national frontiers."[2] What I wish to contend, rather, is that the idea and materiality of the "national frontier," premised on a notion of the nation as a territorially and demographically circumscribed entity, is not prior to but takes shape within the context of racialized migration. A blurring of the vocabularies of nationality and race is a founding strategy of the modern nation-state that makes it impossible to inquire into the modern state without attending to its creation in a global context of colonialism and racism. The passport is one concrete technology that harnesses this strategy to produce the "nationalized" migrant body. The historical contours of the regulation of racialized migration, which impelled a state monopoly over migration practices, produced the specifically modern imbrication of the state, the nation, and race. Here "race" is understood as a "national attribute" that is codified in the state document of the passport. To suggest that the passport is a technology that nationalizes bodies along racial lines is, therefore, to track the itinerary of a process of subject constitution where both terms, nation and state, are implicated in discourses of race.

My second argument is concerned with historicizing state sovereignty, particularly in relation to migration. As we saw in chapter 1, the initial state interventions in monitoring the movement of indentured labor from India occurred amid challenges to the authority and legality of the state in regulating the movement of "free" subjects, especially between British colonies. These challenges were answered on two grounds: the first stemmed from the paternalism of the state, which felt it "could not divest [itself] of the interest which [it] felt bound to take in the well-being of those who might be tempted to try their fortune by engaging as labourers in other countries."[3] Though conceding "that this practice [had] no foundation in any existing law," the regulations were said to be warranted in order to ensure the security and well-being of the laborers, especially given the "necessary ignorance" of the "class of persons so engaging themselves."[4] Second, the regulation of "free" subjects was defended on the grounds that "it is a distinction common to every metropolis, that their colonies are governed . . . by special laws, because the elements of society are not the same therein as in Europe."[5] The peculiar situation of the colony not only justified the differential application of the law and made the term *British subject* itself "susceptible of important division and modification"; it also, as we saw, effected a transformation in understandings of state sovereignty in relation to migration.[6] The rule of colonial difference—the paradoxical and confounding principle of the non-universal applicability of universal principles—that underwrote these transformations would, in the early twentieth century, once again come to structure the international system of states seeking to regulate the movement of "free" subjects. An analysis of the events, debates, and conundrums that attended the proposal to restrict Indian migration to Canada via a system of passports shows how its eventual acceptance was effected through a novel understanding of state sovereignty and security made on national grounds. Working against solely diffusionist or mimetic understandings of modern state formation and sovereignty doctrine, I explore the constitutive relationship between colonialism and state sovereignty to demonstrate the contingent, eventful nationalization of migration, perhaps best encapsulated in the passport.

Pesky Precedents: The Legislation on Indian Emigration

The primary concern animating the nineteenth-century regulations was to facilitate movement and establish a series of criteria by which the migration of indentured Indian labor could be construed as "free," and thereby guard against charges that indenture was but slavery by another name. At the center of the regulations, which served to install the juridical category of "free labor," was a renovated contract where nominal consent signified agency and freedom. As I have noted earlier, under the indenture system,

Indians moved to a number of colonies—Mauritius, Trinidad, Jamaica, Guyana, Fiji, Tanzania, Kenya, Uganda, South Africa, and several other destinations. However, until the early twentieth century the state monitored only the large-scale movement of indentured Indian labor and did not interfere with the migration of those not participating in the state-controlled indenture system. As demonstrated in chapter 1, the regulation of indentured migration was instituted as an *exception* to the overarching principle of free movement. Indeed, within the law, the terms *emigrate, emigration*, and *emigrant* referred *only* to indentured labor. Thus, Act XXI of 1883, the definitive Indian emigration legislation before 1917, states: "'Emigrate' and 'Emigration' denote the departure by sea out of British India of a native of India under an agreement to labour for hire in some country beyond the limits of India other than the island of Ceylon or the Straits Settlements."[7]

The term *to labour*, in turn, had been interpreted as "manual labor," thus effectively exempting emigrants from the wealthier classes. Further, the act specified, expressly, the countries to which one could "emigrate." Thus, "emigration" referred *only* to indentured migration to specific destinations such as Mauritius, some countries in the Caribbean, South Africa, Fiji, etc., and expressly excluded Ceylon (Sri Lanka) and the Straits Settlements (which included present-day Malaysia and Singapore). This circumscribed definition of "emigration" had been the source of substantial ambiguity and confusion over the years, leading to a series of reinterpretations and redefinitions, although not to legislative changes. For instance, an 1862 proposal to send 150 Indians for the police force in Hong Kong resulted in limiting the term *to labour* to refer only to manual labor.[8] The recruits could thus be sent to Hong Kong, to which "emigration," as understood within the legislation, was not permitted. In 1899, the transfer of four hundred Indian soldiers to Mombassa prompted the Government of Bombay to ask if the soldiers should be designated as "emigrants" and the request of the military declined.[9] Once again, the executive took the position that while already under a contract to labor the soldiers could not be considered "emigrants" since "the work of a soldier can scarcely be defined as 'manual labour'; and if this view is accepted, it is clear that the terms 'Emigrate' and 'Emigration' as defined by the Act [XXI of 1883] do not apply to the case of these soldiers."[10]

Given the nature of the legislation, emigrants who did not contract to labor prior to embarking on their journeys, or those not engaged in "manual labor," or those whose destination was not among the specified countries of "emigration," could travel, unhindered, especially between parts of the British empire. In other words, thus far, the empire-state did not exercise a monopoly over the mobility of people.

Within the history of Indian migration regulations, it was only with the increased migration of non-indentured—or "more free, 'free' migrants"— to several white-settler colonies (such as South Africa, Australia, Canada, Argentina, and the United States) in the first decades of the twentieth century that we witness vigorous demands to extend state control to cover all types of movement. The explicit aim of these demands, to restrict and prohibit the migration on racial grounds, was of a piece with broader anti-Asiatic sentiments, policy, and legislation across these jurisdictions that conceived various settler colonies as racially white. Though there was a similarity across states in the demand for white-only settlement, the legal and other mechanisms mobilized to achieve this outcome, their levels of success, and their impact varied wildly. Thus, racial discrimination in immigration was implemented through a host of mechanisms that emerged from the unique circumstances of each case and included varied innovations: marriage regulations, the imposition of a head tax, the prescription for education/literacy tests, specifications regarding identity documents, precise regulations concerning the trajectory of voyages, and "gentleman's agreements" of compromises between states on imposing restrictions on emigration. The mechanisms deployed were occasioned by particular social, political, and economic conditions, which spoke to and utilized differing—sometimes conflicting—legal logics and justifications. We have already seen, in the previous chapter, how South Africa, after having suspended indentured migration, mobilized objections to polygamous marriage as a mechanism to curtail non-indentured Indian migration. The particular mechanisms mobilized in Canada to restrict Indian migration through a system of passports are especially important since they both encapsulated and helped precipitate a more widespread change in the legal, we might even say the doctrinal, thinking on migration control.[11]

Indian Emigration and the Formation of the Passport:
The Canadian Connection

The anxiety over the migration of Indians to Canada began in earnest in late 1905, with the arrival of about two thousand Indian men at Vancouver. The anxiety is evident in the actual content of the correspondence and in what was to become an intense, frantic communication, primarily via telegrams and a series of confidential memos and reports, between the clunky state triangulation of Canada (a self-governing British Dominion), Britain (the seat of imperial power), and India (a non-self-governing colony, populated, nonetheless, by British subjects).[12] The arrival in Vancouver of "some 2,000 people from Northern India" prompted the Governor General of Canada to telegram the Secretary of State for the Colonies in London stating that the men had "doubtless come under misrepresentation as they are not suited to climate, and there is not sufficient field for their employment. Many in danger of becoming public charge and thus subject to deportation under law of Canada."[13]

The three points raised by the Governor General in this brief telegram, concerning the climate, the alleged lack of labor demand, and the possibility of destitution, would quickly find their way into a five-point memorandum issued by the Government of Canada on November 2, 1906 as the bona fide reasons for discouraging Indian immigration to Canada.[14] Two of the five points directly cited the climate as the chief reason to restrict movement since the "transfer of any people from a tropical climate to a northern one . . . must of necessity result in much physical suffering and danger to health."[15] In addition, the memorandum claimed that the "caste system which is universal among these people is seriously in the way of their employment," and "the work, for which they are required is necessarily rough and hard, and not of a character . . . for which they are physically fitted."[16] The memorandum therefore concluded that "should the immigration continue, large numbers must become a [public] charge . . . in which case they would be subject to deportation under Canadian immigration laws."[17]

The memorandum made no direct mention of a biological notion of race to prohibit Indian immigration, but appealed, instead, to the cultural and climatic incompatibility of Indians to the Canadian environment. We see here, with exceptional clarity, the deployment of what Paul Gilroy identifies as the culturalist trope of racism and what Étienne Balibar calls a "differential racism." As Balibar describes it, "differential racism" purports to function within "a framework of 'racism without races' . . . whose dominant theme is not biological heredity but the insurmountability of cultural differences."[18] Culturalist or differential racism is frequently understood as emerging in the era of decolonization, which, at least implicitly, pertains fundamentally to transformations in state power. Here, I situate its emergence earlier and suggest that it is crosshatched, on one hand, with the development of nationalist movements in the colonized world and, on the other, with the progressive solidification of a notion of the "national frontier," which served to congeal boundaries around territories as much as it did around populations. Indeed, culturalist racism succeeds, precisely, in securing an identity between people and territory such that both come to be described as "national."

Even as the memorandum claiming an insufficient labor demand in British Columbia and obsessed with the climatic and cultural unsuitability of Indian emigrants to Canada was drafted and circulated, N. D. Daru, an Indian official attached to the Geological Survey of Canada, offered an alternative account, writing that there "is no doubt that British Columbia is greatly in need of labour and that the Indians who come in, are readily taken up by employers . . . So many employers asked me to enable them to get these men that if I had over a hundred more, I could easily have placed them."[19] Similarly, Colonel Falk Warren, an artillery officer in British Columbia, immersed in the ideology of empire, would write:

> There are now between 2,200 and 2,500 [Indians] in the province of British Columbia all come from the Punjab; they are mostly Sikhs. None are allowed to land who are destitute, so all have some ready money in hand, some as much as thirty pounds or more. . . . They are a stout and able bodied set of men. A large proportion are ex-soldiers. Their conduct has been exemplary. I am not aware that a single one has been convicted of any crime, however slight. . . . The fact of their coming to this country shows the enterprise and daring of the men as also their trust in seeking work in countries under the British flag at such a distance from their own homes.[20]

The memorandum's concerns about a lack of employment and resulting destitution were unfounded. In January 1907, "of the 2,200 men there are not more than 50 to 60 out of work . . . [and the] public funds have not been called upon to expend anything upon any of these men."[21] Indeed, Lord Grey, the Governor General of Canada, was himself forced to agree with the special confidential assessment of the situation by Colonel Swayne, Governor of British Honduras:[22] "[According to Colonel Swayne] the position of the Sikhs and Hindus at present in British Columbia leaves little to be desired. He [Colonel Swayne] found no justification for the statement that large numbers were unemployed or in distress . . . [and] that the climatic conditions in British Columbia are not so severe as those to which the Sikhs are accustomed in their native country . . . [thus] it will be impossible in future to urge climatic considerations as a reason for discouraging, on humanitarian grounds, the emigration of Sikhs to British Columbia."[23]

In an attempt to curb the migration, the Canadian state had resorted to arguments regarding the severity of the climate and the constant unsubstantiated panic about impending, large-scale destitution due to the inapplicability of Indian Emigration Act XXI of 1883. Since the Indians in Canada were not already under contracts to labor, they did not count as emigrants under the act, which could thus not be enforced to prevent migration. The *emi*gration, in other words, was deemed "free" such that the Government of India could not, within the prevailing legislation, control it.[24] The Government of Canada, for its part, was constrained in drafting restrictive *immi*gration legislation that specifically targeted Indians since this would have exposed, in an indubitable way, that notwithstanding citizenship of/subjecthood in the Empire, different British subjects enjoyed differential access to mobility. The involvement of the state in restricting migration *at either end* would have exposed the rule of colonial difference in a particularly salient and unacceptable manner.

In this context the Government of Canada suggested, in 1907, the implementation of a system of passports, issued selectively, to curtail Indian migration. Prime Minister Wilfrid Laurier offered two options for the consideration of the Colonial Office in London and the Government of India. The first was a monetary requirement of $200 for each immigrant seeking entry to Canada. This was deemed "necessary to avert real suffering and distress and consequently would appear to us to be called for in the best interests of humanity."[25] The second option was the adoption of a system of passports that met three conditions: "(1) prohibit Hindoos from going to Canada without passports, (2) to limit the number of passports issued to a number agreed upon by the Governments of Canada and India, and (3) to request Government of Canada to deport all Hindoos arriving at Canadian ports without passports."[26] Laurier's suggestions did not represent an innovation in Canadian migration control. As tactics of prohibition, Canada already imposed a "head tax" on Chinese migrants and had concluded a "gentleman's agreement" with Japan, whereby the latter would restrict the number of emigrants to 400 annually (via a system of passports).[27] The passport here functioned less as a document of identity and more as what we now know as a quota system for visas.[28] Laurier's attempt to utilize these preexisting strategies of prohibition with regard to Indians met with mixed success. On the one hand, the Viceroy of India rejected the passport system, writing in a telegram:

> We recognize peculiar difficulties of Canadian Government and appreciate the conciliatory attitude with which it has approached this difficult question, but after very careful consideration, regret we are unable to agree to any proposal for placing in India restrictions such as are suggested on emigration of free Indians or to suggest any further action on our part to check it. Any such measure would be opposed to our accepted policy: and it is not permissible under Indian Emigration Act XXI of 1883. . . . In present state of public feeling in India we consider legislation of this kind to be particularly inadvisable.[29]

In 1908, the "present state of public feeling in India" was becoming deeply anti-colonial and nationalist. Indeed, in that year the Indian National Congress adopted *swaraj* (self-rule), "like that of the United Kingdom," as its goal and put enormous pressure on the state to concede at least partial self-government.[30] Alongside the more moderate, "constitutional" approach of the Congress, there emerged numerous other radical and revolutionary groups opposing colonial rule. The situation of Indian emigrants in different parts of the Empire added to the ferment and nationalist demands within India. As we have seen, the most prominent of these were the agitations of Indians in South Africa that not only had become a troublesome matter in South Africa but had also led to uproar in India.[31] Unlike most other colonies, particularly white-settler colonies, South Africa was unique in having a large population of both indentured and "free" Indian immigrants. In general, the demands of Indian nationalists in both South Africa and India aimed to end indentured emigration, but restrictions on "free" emigration incited widespread opposition. As emigration became a more central element in Indian politics and anti-colonial demands, the Government of India wanted to avoid situations such as legislative measures that expressly discriminated against Indians *qua* Indian. Thus, while, on the one hand, the Viceroy rejected Laurier's passport proposal, on the other he suggested that Canada instead pursue suitably disguised methods of discrimination to curtail the immigration. For instance, they could "require certain qualifications such as physical fitness . . . and the possession of a certain amount of money."[32]

Simultaneous with the sensitive situation in India, the "public feeling" in Canada was becoming overtly racist toward "Asiatics" and "Orientals." Where the official authorities cited "humanitarian" considerations of climate and labor demand as reasons to restrict the immigration, both N. D. Daru and Falk Warren pointed to "anti-Asiatic sentiment" as the problem needing attention. The immigration, they said, was opposed by "the whole labour element of this country," who had engineered a campaign of "calumny and vituperation" against the Indians; by the press, which had "not merely taken up a hostile attitude [toward the immigrants], but [had] not scrupled to publish the rankest falsehoods about the Indians"; and by organized Anti-Asiatic Leagues in Canada and the United States of America.[33] In fact, on September 6 and 7, 1907, Anti-Asiatic Leagues had orchestrated riots in Bellingham, Washington, and Vancouver, British Columbia, against Indians and "Asiatics." The Bellingham riots had caused some four hundred Indians to leave the United States and move to Canada "seeking the protection of the British Crown."[34] The Canadian state thus found itself in an odd position: attempting, on the one hand, to restrict the immigration of Indians and being bound, on the other hand, to provide refuge for Indians due to their mutual membership in Empire. Racism was clearly operating in numerous registers, from the culturalist racism of the state to the physically violent racism of the rioters. And it was race articulated to a space, increasingly described as "national," that would subtend subsequent immigration regulations, including the emergence of the passport.

The entire machinery of Empire, from the Government of Canada and the Government of Hong Kong, to the different district authorities of the Government of India, was enlisted to inquire into every aspect of the migration. A report on the character of "The Hindus" was prepared by no less an authority than the Minister of the Interior of Canada; a secret agent was employed to infiltrate the Anti-Asiatic League of Canada in order to determine their support base and funding source; authorities in Hong Kong were directed to provide information on every ship that sailed, including details of the number of Indians on board and their financial situation; reports on the factors encouraging the migration were elicited from the Government of India; ethnographies of the immigrants themselves were conducted to understand their motivations; the role of shipping companies in assisting the traffic was assessed; Mackenzie King, then Deputy Minister for Labor (later a long-serving Prime Minister of Canada), was dispatched for secret consultations with the colonial government in London on "the subject of immigration from the Orient and the immigration from India in particular."[35] In short, what the relatively insignificant migration of two thousand Indians to Canada instigated was the eruption of a variety mechanisms for generating, obtaining, and collating knowledge on every aspect of the movement of Indians to Canada.

This process of knowledge production was impelled by a complex conjuncture of ideologies of racism. At the same time, the mechanisms and techniques employed in the service of this ideological conjuncture are specific to the modes of governance and the administrative machinery at the disposal of what we call the modern state. It was not just that the state was propelled into action by racist ideologies or that race structured state policy, but that such moments functioned as catalysts that aided the development of a specifically modern regime of state power. Unlike early colonialist ideology that worked on the classic colonialist framework of the "civilizing mission," the kind of racism deployed within these twentieth-century immigration regulations is distinguished by the development of the culturalist trope of racism and by the deployment of a series of what Bernard Cohn calls "investigative modalities." An investigative modality, as Cohn explains, "includes the definition of a body of information that is needed, the procedures by which appropriate knowledge is gathered, its ordering and classification, and then how it is transformed into usable forms such as published reports, statistical returns, histories, gazetteers, legal codes, and encyclopedias."[36] In more Foucauldian terms, what we see in the process set in motion in a bid to restrict the immigration of Indians to Canada is the critical role of knowledge in the consolidation of a form of power endemic to the modern state.

The knowledge amassed through the use of multiple investigative modalities yielded a small, yet crucial, piece of information: authorities learned that a proportion of Indian immigrants to Canada were re-immigrants who had worked in a country other than India and went to Canada having heard that livelihoods, if not fortunes, could be made there. Otherwise, the bulk of the emigrants came from Punjab and, in the absence of the possibility of making a direct voyage from India to Canada, went first from Calcutta (Kolkata) to Hong Kong and thence to Canada. With this detailed and minute information in hand, on January 8, 1908, the province of British Columbia passed an ingenious and notorious Order-in-Council, which stated that "immigrants shall be prohibited landing [in Canada], unless they come from [their] country of birth or citizenship by continuous journey, and on through tickets purchased before starting."[37] This order effectively prevented both re-immigrant Indians and immigrants coming directly from India to enter Canada: the former since they did not come from what was deemed their "country of birth or citizenship," and the latter since the "continuous journey" condition was literally impossible to fulfill, as no steamship company operated a direct transit from any port in India to Canada. Companies, such as the Canada Pacific Railway Company, that wished to extend their business into passenger shipping were pressured not to sell "through tickets" to Indians.[38] In this way, private businesses were also incorporated into an increasingly dense, if covert, web of active racial discrimination.

The precise wording of the initial continuous journey regulation is significant for what it reveals about law making and implementation. Directly following on the announcement of the regulation, a Russian and a Frenchman, "both well educated men apparently, one an electrical engineer and the other a bank clerk," arrived at Vancouver not by "continuous journey" from their "country of birth or citizenship," but rather from Japan.[39] Under the regulation of January 8, both were disallowed entry into Canada. They would have been deported had it not been for the intervention of U.S. immigration officials at Vancouver who were "glad to pass them on into the United States."[40] Recounting the incident to Prime Minister Wilfrid Laurier, the secret agent T. R. E. McInnes would comment: "I, of course, have no status to advice in such a matter, but I know that the Regulation was never intended to be enforced in this absurd manner."[41] He recommended that the regulation be reworded to state that immigrants "*may* be prohibited [landing]—not *shall* be."[42] The aim of the regulation was specifically to prevent the entry of Indians into Canada. Its original phraseology, combined with the bureaucratic logic of the state functionaries, led to the unintended consequence of it being implemented in an "absurd manner." The Ordinance was immediately reworded so as to enable officers to use discretion and thereby permit the entry of white immigrants, regardless of where they embarked on their journeys.

The correspondence surrounding the regulation enables us to see the machinations involved in interweaving racism and a juridical liberalism. The efficacy of the unnamed racist strategy of the law was dependent on bureaucratic discretion, with the exercise of that discretion crucial to maintaining the spirit, as opposed to the letter, of the law. The introduction of caveats and the necessary bureaucratic discretion they entail has become an increasingly important part of current legal regimes, particularly in domains such as migration law. Now an entire branch of law, called administrative law, it serves as a way of incorporating exceptions in and to the rule and shielding decisions on such exceptions from judicial review.[43] As with several other contemporaneous legal measures undertaken by different white-settler states, the continuous journey regulation embodied the "open secret" of racial thinking in the law: without knowledge of the very particular contexts, their racist rationale was obscured. Moreover, once admitted into legislation, the courts and judges largely committed themselves to simply implementing the law, rather than interrogating either its motivations or its legitimacy. The latter possibility, however, always remained, haunting the legitimacy and efficacy of the law.[44]

In addition to the continuous journey regulation, Canada also actively pursued the suggestion of the Government of India, mentioned earlier, that it follow strategies of selective racial discrimination without naming race as such. Thus, along with the continuous journey regulation, it imposed on Indian immigrants the $200 monetary requirement. The twofold rationale offered for these restrictions, let us recall, was the purported inability of people from a "tropical climate" to "readily adapt themselves" to the Canadian climate coupled with the claim that the migration "would result in a serious disturbance to industrial and economic conditions in portions of the Dominion, and especially in the Province of British Columbia."[45] Hence, an effective restriction on the immigration of Indians was deemed desirable "not less in the interest of the East Indians themselves than in the interest of the Canadian people."[46]

Together, the continuous journey stipulation and the monetary requirement made the entry of Indians into Canada virtually impossible. In fact, between 1909 and 1913, only twenty-seven Indians managed to enter Canada.[47] These immigrants successfully established that they were returning immigrants with Canadian domicile. However, the Canadian government continued to press for the adoption of a passport system along the lines of their 1907 proposal.[48] In large part, this was due to the confidential report submitted by Colonel Swayne, Governor of British Honduras. His report stated not only that barring immigration on grounds of climatic considerations was hogwash but also that many employers, such as mill owners, preferred to employ "Sikhs" over white men since they "can be more safely relied upon to give continuous employment."[49] The cause for anxiety now was the possibility of Indian immigrants making and saving substantial sums of money, despite the "hostility of the white trade unions."[50] Swayne speculated that the return of "Sikhs" "enriched with the savings of five years' employment in British Columbia would have a disquieting effect on the Punjab population."[51]

The "disquieting effect" Swayne referred to concerned the agitation that would ensue with the realization that the climatic rationale was unfounded and that the returning immigrants were in a position to provide new emigrants with the funds necessary to meet the monetary requirement. Alongside the problems Swayne identified, the Canadian government was aware that the continuous journey regulation constituted a crude, stop-gap measure whose efficacy could crumble for a number of reasons, and that it was insufficient to guard against a fresh attempt at migration.[52] This prompted their claim that "unless some further regulations restricting the influx can be established, there may be a danger of new developments arising with results possibly prejudicial to British interests *in* India."[53] If, moreover, the Government of India did not impose restrictions on emigration, then "the only alternative would appear to be the adoption of further, and from an Imperial point of view undesirable, regulations restricting British Indian emigration to Canada by His Majesty's Canadian Government."[54]

Despite the veiled threats of imminent doom, the Government of India refused to acquiesce to adopting the passport system. Conceding that the return of Sikhs "enriched with five years' employment" and the "knowledge that Indian labourers are preferred to white men, may give a fresh impetus to emigration to Canada," the Government of India maintained that they could not be party to policy or legislative measures that would strengthen the continuous journey impediment: "No new arguments have been adduced which would justify us in abandoning the position we have so far adopted on this subject, and, while we are prepared to take the necessary steps to warn intending emigrants of possible trouble in the country to which they are bound, we cannot propose any measures, such as the suggested introduction of a system of passports, which would *publicly* identify us with the policy of exclusion of Indians from other portions of the Empire."[55]

The Unmaking of the Empire-State

A number of fundamental issues were at stake in the passport proposal and the Canadian regulations more generally. Not least of these was the very foundation and legitimacy of the British Empire that had come to rest crucially on the definition of the term *British subject*. The Governor General of Canada was entirely correct in remarking that the imposition of further restrictions was "from an Imperial point of view undesirable." Indians in Canada had sent a petition to the colonial government questioning both the monetary requirement and the continuous journey regulation and demanding their "rights as British subjects with all the emphasis it can command":

> The present Dominion Immigration Laws are quite inconsistent to the Imperial policy because they discriminate against the people of India who are British subjects; as they are forced to produce a sum of $200 before landing, whereas other British subjects are not. . . . [We also] bring to your notice that no such discriminatory laws . . . [exist] against us in foreign countries . . . to whom we do not owe any allegiance whatsoever. Under these circumstances, we most respectfully implore a favourable consideration and prompt amendment to the unfair laws which impress upon us that we enjoy better privileges under foreign flags than those under the British flag.[56]

The petition further claimed that "as long as we are British subjects any British territory is the land of our citizenship."[57] Though put forth partially as an *appeal* to the state, the tacit question addressed the *legitimacy* of the empire-state and its guarantee to ensure the equality of its subjects.

If the matter of being a "British subject" by claiming that "any British territory is the land of our citizenship" was exercising the minds of Indians in Canada, it was at the forefront in the official correspondence I have been detailing. The repeated refusals of the Government of India to support a passport system concerned this very fact. The conundrum to be overcome was how to distinguish between British subjects, members of a single, expansionist state, without calling the entire edifice of the empire into question. It was a sense of empire and *not* of a territorially circumscribed nation that was paramount here. As discussed in the introduction, developmentalist and diffusionist accounts of the historical formation of nation-states frequently fall prey to the presentism that shapes methodological nationalism and methodological statism. Often relying on Westphalian and Weberian propositions, the state is assumed to be, *by definition*, territorially circumscribed and not territorially expansionist. Since European nation-states developed simultaneously as empires, then we must reconsider both the received history of the concept of the nation-state as a territorially circumscribed entity and the received understanding of the notion of sovereignty that accrues to the state via the nation.

This tension between maintaining empire and nation, simultaneously, is evident in the history of Indian migration to Canada. Indeed, control over migration has a crucial bearing on the emergence of the nation-state and is intimately related to the colonial genealogy of the modern state. The Government of India had, as I have already indicated, encouraged the strategy of using race without naming it as nationality, to avoid situations that would "*publicly* identify" them with discriminatory policies and commit them to "incurring all the odium of passing restrictive legislation."[58] The continuous journey regulation had been enacted in this spirit and further regulations modeled on this schema would become a staple of Canadian legislative measures. Witness, for instance, the 1912 correspondence when the Province of Saskatchewan proposed "An Act to prevent the employment of female labour in certain capacities." The act stated: "No person shall employ in any capacity any white woman or girl or permit any white woman or girl to reside or lodge in or to work in, save as a *bona fide* customer in a public apartment thereof only, to frequent any restaurant, laundry or other place of business or amusement owned, kept or managed by any Japanese, Chinaman or other Oriental person."[59] The explicit equivalence in this act of nationality and differential treatment led to frantic confidential communication. Writing to the Governor General of Canada, the Secretary of State for the Colonies stated that while the London office had no objection "to the substance of the legislation which they presume has been rendered desirable by local circumstances" it was "strongly opposed to any discrimination affecting Japanese subjects by name . . . and the Government of India have no less strong objection to any provisions so worded as to discriminate by name or by inevitable implication against natives of British India."[60] Thus, the Government of Saskatchewan should be directed to "amend the Act . . . in such a manner as to remove any discrimination by name against Japanese or British Indian subjects."[61] The overt discrimination *by name* against Chinese subjects was deemed permissible due to "the absence of Treaty engagements with China" and to the "many precedents of Acts differentially affecting Chinese."[62] Thus, "as far as Chinese are concerned . . . it is not necessary to take exception to the terms of the Act."[63] The Colonial Office suggested the act be reworded to state that "no person without a license to be obtained from some executive authority" would be permitted to employ any white woman or girl and "thus by differential treatment in the matter of the grant of licenses, the Japanese and other Orientals could be refused licenses . . . [at] the absolute discretion of the Authority selected."[64]

This event crystallized several issues that merit mention. First, while the regulation was motivated by the urge to protect whiteness, these fears were transferred onto the body of the white woman to be protected by the chivalrous state. The protectionist stance of the patriarchal state was a conduit for an articulation between race and gender.[65] Second, as with the continuous journey regulation, the correspondence surrounding the act in Saskatchewan reveals, once again, how the efficacy of the unnamed racist strategy of the law was wholly dependent on the individual executive authority selected to grant licenses. It was not that individuals inadvertently employed the law in racist ways but that individuals were required to make race-based distinctions to ensure that the law operated in racist ways. Here, ironically, racism could function only by adhering to "the spirit" as opposed to "the letter" of the law. Racism was thus instituted by bureaucratic discretion and the exercise of discretion was crucial to fulfilling the spirit of the law.[66]

Third, the correspondence and, notably, the subsequent acceptance of the suggested amendment to the act point to the dangers of assenting to a teleological narrative of chronologically diminishing racism, a narrative that works in tandem with an appeal to a hermetically sealed notion of "the spirit of the times." What we see in the exchange over the "Act to restrict the employment of female labour in certain capacities" is *not* that racism was so pervasive as to evade interrogation but quite the opposite: that the racist state developed in cognizance of its racism. The dilemmas of racism were evident both in the demand for confidentiality of the documents and in the general policy of not *naming* race. As such, the view that the practice of virulent racism is somehow conducted "unintentionally" or with "the best of intentions" and diminishes over time is not only empirically weak, but lends credence to a progressivist idea of history that deflects attention from the current prevalence of racist ideologies and practices.

The fourth point this event illuminates concerns the particularities of the linkages between nationality, the state, and race. Since, in the (mistaken) view of the Secretary of State for the Colonies, Britain did not have any treaty agreements with China, it was not under any obligation to avoid race-based policies vis-à-vis the Chinese. In this case, race-based policies could be easily rationalized and explained away, post hoc, as the logic of the state and as unrelated to race. The matter was somewhat more complicated with regard to Japan, with whom Britain and Canada did have treaty agreements that could not be jeopardized. In the case of India, the situation constituted a real dilemma, due to common membership in Empire and, hence, common imperial citizenship. Significantly, the strategy of the empire-state, to employ racial difference without naming race, produced the peculiar, though now naturalized, outcome of splitting a racist discourse that was not only able, but forced, to distinguish between the Chinese, the Japanese, and the Indian. This split racist discourse generates nationality not on a terrain of affective bonds of national community but on a terrain overdetermined by race and the relationship between states. Unlike the events in South Africa, where Indian protests mobilized nationalist narratives of community, authenticity, and religious identity, here nationality (i.e., the Chinese, the Japanese, and the Indian) was activated as a *state assignation* and was an outcome of how the state rationalized race and implemented a racist agenda. However, this far, the correspondence around migration to Canada contained few appeals to nationalist narratives that bequeath and legitimate nationality on, quite literally, the soil of the territorial nation, which is quite different from nationality as a state designation made on juridical grounds. This was true for all participants in the conversation—the Canadians, the British, the Indians (both the government and British Indian subjects), the Japanese, and the Chinese. This notion of nationality, whereby people are tethered to a geographical space described as a nation, articulated to other discourses would eventually culminate in the passport as the definitive state document authorizing *national* identity and further curtailing mobility.

The Canadian restrictions vis-à-vis Indians had not only prevented the further migration of Indians into Canada, they had effectively functioned as a mechanism causing the *exodus* of Indians already there, since their families were unable to join them.[67] Thus, from 6,000 men in 1906, the number of Indians in Canada fell to 4,500 (of whom only three were women) by 1915.[68] The regulations that were implemented, as we have seen, attempted to retain the appearance of an unfissured imperial citizenship and keep intact the status of "British subject" even as the actual motivations and effects of these regulations were quite the opposite. As the intricacies involved in using a racist strategy without naming race were becoming increasingly complex and impossible to disguise, the difficulties with retaining the juridical appearance of an indivisible, unitary "British subject" were also exacerbated. There are, in other words, only so many convolutions one may enact around race (understood here as crude physiognomy) without naming it.

Indians in Canada had, of course, seen through these thinly veiled strategies of racial exclusion and had vehemently and repeatedly protested the differential treatment they received. In fact, they had gone further. Among the Indian emigrants in Canada were several who had been actively involved in revolutionary politics in India and many had left India in order to escape arrest for "terrorist" activities.[69] The temples, schools, and associations established by the Indian community in Canada also functioned as covers for "seditious" activity, including the publication of several newspapers and pamphlets that were circulated in North America and sent to India.[70] William Charles Hopkinson, the translation officer attached to the Canadian Immigration Office in Vancouver, simultaneously gathered information on "seditious" Indians and was deeply enmeshed in a web of surveillance and intelligence activity on behalf of the governments of Canada, India, and the United States. The aim of the revolutionaries was unambiguous: to overthrow British rule in India, if necessary by violent means.

Hence, the revolutionaries utilized every instance of differential treatment toward Indians to undermine the British empire and whip up support for their cause—not only in Canada but, perhaps more significantly, in India.

In September 1913, the Secretary of State for India in London once again asked the Government of India to reconsider the Canadian passport proposal and thereby restrict Indian emigrants to a pre-agreed number. Yet again, officials in India found Canadian anxiety misplaced, viewing the continuous journey regulation as an adequate barrier to unfettered migration into Canada. Dismissing the possibility of a direct steamship from Calcutta to Vancouver as purely "hypothetical," S. H. Slater, Undersecretary to the Government of India, would write: "We should decline to be drawn into a statement of what we would do in hypothetical circumstances."[71] W. H. Clark concurred: "Apart from any other consideration, it would be a most reckless thing for us to commit ourselves to incurring all the odium of passing restrictive legislation at this juncture merely in view of the possibility of a through steamer communication being established with certain hypothetical results."[72] Thus, until September 1913, the Government of India was steadfast in its refusal to participate in a system to restrict emigration to Canada grounded in the principle of "complete freedom for all British subjects to transfer themselves from one part of His Majesty's dominions to another," with S. H. Slater stating, "We have consistently declined to be parties to such a policy, and there seems no reason why we should abandon our attitude."[73] Reiterating their position from 1908 concerning the "state of public feeling in India," the Government of India warned that "the state of public feeling now [i.e., in September 1913] render any such legislation even more undesirable . . . if we attempted it, we should raise a storm of protest all over India; and without legislation we have no power to restrict free emigration."[74] By 1913, the situation of Indian emigrants in Canada and South Africa had become something of a political cause célèbre within India.

Nationalization as Event

Two related events caused a radical alteration in the position of the Government of India on the Canadian passport proposal. In October 1913, fifty-six Indians arrived in Victoria, British Columbia, aboard the *Panama Maru*. All claimed prior domicile as the basis for (re)admission into Canada. William Hopkinson, the translator, who doubled as an intelligence agent, let in seventeen whom he thought he recognized. The remaining thirty-nine immigrants were denied admission on the grounds that they had violated the Orders-in-Council. One of them, however, escaped from the Immigration Hall where they were locked up.[75] The Indian community managed to challenge the decision of the Board of Inquiry of the Immigration Department by demonstrating that the language of the Orders-in-Council, which had been cited to prohibit the thirty-nine immigrants from entry into Canada, did not conform to the language of the Canadian Immigration Act of 1910. In other words, since the Orders-in-Council had been used as the basis for denying admission to the Indians, and they were not in consonance with the Immigration Act, the decision of the Board of Inquiry was rendered void.[76] Thus, thirty-four of the remaining thirty-eight immigrants were allowed entry. Four were denied admission on medical grounds but they too succeeded in running away from the Immigration Hall.[77] In sum, all fifty-six Indians who had arrived on the *Panama Maru* managed, in one way or another, to enter Canada.

The news of the court victory in the case of the passengers of the *Panama Maru* soon spread, and this, combined with the active encouragement of revolutionaries and others in Canada, provided the impetus for Sardar Gurdit Singh (a businessman and entrepreneur) to hire a ship, the *Komagata Maru*, to sail from Hong Kong to Vancouver, with stops at Shanghai, China and at Moji and Yokohama, Japan. In all, the ship gathered 376 passengers, mostly Sikhs, and arrived off the coast of Vancouver on May 23, 1914.[78] The Canadian government had, by now, eliminated the dissonance between the Orders-in-Council and the Immigration Act that had allowed the passengers of the *Panama Maru* entry into Canada.[79] However, there were still grave risks to testing, in court, the absurdity of the continuous journey regulation or the sticky issues regarding the definition and legal entitlements of a British subject.[80] Though the *Komagata Maru* had not sailed by continuous journey from India, its arrival indicated, in no uncertain terms, the practical fragility of the continuous journey regulation as providing a strong bulwark against Indian migration. If not via the establishment of a regular steamship service, the audacious journey executed by Gurdit Singh brought into the realm of the plausible that a ship could be chartered in India to sail to Canada. What Slater and Clark of the Government of India had dismissed as the "hypothetical" imaginings of Canadian officials was now a real—and threatening—possibility. The *Komagata Maru* made apparent to officialdom that stronger measures than those embodied in the continuous journey regulation and the monetary requirement of $200, for each migrant, needed to be devised to effectively prevent Indian migration.[81] Moreover, the arrival of the *Komagata Maru* caused a furor in the Canadian House of Commons and in India. Since outright appeals to biological racism and racial superiority could not be legally countenanced, much of the debate in Canada again substituted cultural racism for biological racism while advocating for the principle of the sovereignty of states based on *national* grounds. For instance, Frank Oliver—who had been Minister of the Interior at the time of such measures as the continuous journey regulation, but now formed part of the opposition—voiced his objection to the immigrants in the following terms:

> The immigration law as it stands is a declaration on the part of this country that Canada is mistress of her own house and takes the authority and responsibility of deciding who shall be admitted to citizenship and the privileges and rights of citizenship within her borders. . . . This is not a labour question; it is not a racial question; it is a question of national dominance and national existence. . . . This [the *Komagata Maru* incident] is an organized movement for the purpose of establishing as a principle the right that the people of India, and not the people of Canada, shall have the say as to who may be accepted as citizens of Canada.[82]

While Frank Oliver attempted to cover over the racist motivations for disallowing Indians into Canada in terms of the threat they posed to the very definition of the sovereignty of the nation, Wilfrid Laurier—former Prime Minister, now also among the opposition—was more direct in his comments: "The people of Canada want to have a white country, and certain of our fellow subjects who are not of the white race want to come to Canada and be admitted to all the rights of Canadian citizenship. . . . These men have been taught by a certain school of politics [i.e., liberalism] that they are the equals of [white] British subjects; unfortunately, they are brought face to face with the hard facts only when it is too late."[83]

Laurier's comments are remarkable for their candor, since the kernel of the entire struggle over the passport system was precisely about how to effect racial exclusion without naming race. While the necessity and validity of racial exclusion was widely accepted in official circles, as is abundantly clear from the confidential correspondence, the overriding discourse of liberalism made it impossible to actually implement policies that directly cited race. We have seen how the different arms of the empire-state, by incorporating caveats into each policy, utilized the strategy of bureaucratic discretion, time and time again, to circumvent this problem. The state functionaries, moreover, were fully aware of this. Thus a confidential memorandum states: "The efficacy of these Acts, in fact, rests to some extent on a subterfuge."[84] To reiterate, the fundamental dilemma was the fact that Indians were British subjects and the premises of the liberal state now made it exceedingly difficult to distinguish between subjects of the British Empire. As with the 1910 petition from Indians in Canada protesting the continuous journey regulation on the basis of membership in Empire, the meetings and memoranda caused by the *Komagata Maru* incident, appealed, once again, to the juridical definition of Indians as British subjects: "The deep loyalty of Indians to the British Raj springs from the consciousness that the British maxim *par eminence* is that of equal justice and fair play. Under the aegis of the British ideal of justice, Indians only seek for equal chances with their fellow-subjects in other parts of the Empire, so that they may legitimately feel the pride of being citizens, in the full sense, of the Great Empire over which the sun never sets."[85] Based thus in notions of justice and fair play, the memoranda also questioned the unrestricted entry of white inhabitants from other parts of the British Empire into India and, further, threatened "that Canada must be made to understand that she is dealing not with 600 [*sic*] men only, but with 33 crores [330 million] of Indians."[86]

The *Komagata Maru* incident precipitated a rapid transformation in policy. In fact, the very same S. H. Slater of the Government of India, who in September 1913 had vehemently rejected the passport system and opposed any restriction on emigration, seven months later voiced a diametrically opposed position, writing that "circumstances are now compelling a stricter definition of such phrases as . . . 'membership of the British Empire.' It is now conceded that such membership does not carry with it the right of free entry to all parts of the Empire. [Therefore,] in this narrower view . . . it will no longer be held that every measure of exclusion of Asiatics from territories forming part of the Empire is necessarily and *ipso facto* an injustice to Indians."[87] Slater's remarks—widely echoed by officials of the Government of India—point, unambiguously, to a radical *change* in legal and legislative thinking. Until 1914–15, state control of indentured migration to specific destinations was the only *exception* to free movement, whose object, let us recall, was precisely, if paradoxically, to affirm the general principle. For over seventy-five years, from about 1834 to the early twentieth century, the principle of unfettered, free movement had shaped the legislation of the Government of India, with state control over indenture directed at facilitating and enabling, not constraining, the movement. The regulation of indentured migration, under the overarching principle of free movement, had decisively shaped the decade-long discussions and dissensions between India and Canada on devising mechanisms to prohibit migration to Canada, without altering the legislative framework. As a consequence of the *Komagata Maru* event the very rationales that organized migration would change. The challenge now was how to give practical and legislative effect to the changed thinking embodied in Slater's remarks. How could prohibition on migration be achieved while averting widespread political protest by Indians across the globe?

The comments of R. W. Gillian, also of the Government of India, offer the clearest statement of the rationale to justify broader migration regulations. Gillian pointed out that the Government of India's policy in rejecting the passport proposal and its reluctance to interfere with "free" migration, though resting on the principle that "a British subject [had] a right to go and reside in any part of the Empire," had a "double aspect."[88] On the one hand, the government had refused to interfere with "free" emigration from India and, on the other, had not balked at suggesting that different parts of Empire impose all manner of restrictions on immigration so long as their racist motivations were suitably disguised. This, wrote Gillian, "is what appears to me an inconsistency. We adhere to our policy, while abandoning the principle on which it has always rested."[89] Moreover, the intransigent position of the Government of India could not, in Gillian's view, "be defended on its merits, since it denies in effect the right of our Colonies and even of other countries to settle their own affairs."[90]

Gillian here echoed Frank Oliver's comments in the Canadian House of Commons that control over immigration is somehow fundamental to the very definition of state sovereignty, a line of reasoning that would soon shepherd all migration under state control. However, as Gillian continued, "If the right of Canada or Australia to manage their own affairs is admitted, what about India? If the right is denied to her, the result is immediately to emphasize her subjection in an extremely unfortunate manner."[91] What was required was a mechanism that would "secure some kind of reciprocity" and "which [would] above all things . . . have the *appearance* of giving equal treatment to British subjects residing in all parts of the Empire."[92] It was imperative to vigorously hold on to inscribing the letter of the law as universal, while ensuring that, in practice, this universality would function differentially.

The sine qua non of a resolution that had the appearance of "reciprocity" was the introduction of a notion of "nationality" into the regime of migration control. Mobilizing a notion of "nationality," as an alibi for race, was amenable to arguments of reciprocity and enabled a covert racism. Nations could now engage in a reciprocal (if asymmetrical) discrimination against other nations, argue that every nation had the "inherent" right to institute such discrimination, and claim that it bore no relation to racial thinking. Encoding this radical transformation in migration control in terms of reciprocity between nations gained traction not only in official thinking. Reciprocity—or, more accurately, a retaliatory sentiment—also shaped nationalist Indian views on migration. For the latter, national honor and prestige could be salvaged only by imposing restrictions on migration into India. Ian Fletcher, in his survey of a range of publications, shows how arguments for reciprocity were central to nationalist demands to address the exclusion of Indians from parts of Empire.[93] Ironically, then, the official and the anti-colonial nationalist views converged. The legal expression of this logic was an interim Reciprocity Resolution, eventually followed by the Reciprocity Act.

In this way, by recourse to the idea of states securing sovereignty through an appeal to the "national," the principle of the "complete freedom for all British subjects to transfer themselves from one part of His Majesty's Dominions to the other" was abandoned, and the category of "British subject" was rendered available for division and differentiation based on the rule of colonial difference. Since, wrote Viceroy Hardinge, "thoughtful people will agree that states and countries have an inherent right to decide whom they will or will not admit within their borders," the solution reached was, in the words of R. E. Enthoven, to "undertake to furnish passports to emigrants entitling them to admission into the Colonies and India respectively. The number of these permits or passports would be limited by agreement."[94] It was simply fortuitous that the First World War was well under way, thus lending credence to calls for state security. Indeed, the official rules for requiring passports from all Indians proceeding outside India appear as the Defence of India (Passport) Rules, which were a subsection of the Defence of India (Criminal Law Amendment) Act of 1915. This act—more about the defense of a "white Canada" than of India—made embarking on a journey from any port in British India without a passport a criminal offense.[95] Significantly, the passport requirement did *not* apply to indentured labor. Moreover, and perhaps unsurprisingly, this piecemeal, exigent formation of migration law ensured that it lacked a coherent, overarching rationale or set of principles that guided and circumscribed law making. Instead, contingent historical circumstances instigated certain legal and legislative measures and produced a hodge-podge formation, including situations where different laws could explicitly conflict with each other.

Conclusion

The legitimation for state sovereignty has historically been made on different grounds. One of the features that distinguishes the nation-state from other state formations is that the state secures sovereignty via an appeal to the nation.[96] The acceptance of the passport system not only was underwritten by this barely emergent understanding of the sovereignty of states premised on the national, but was central to ensuring its effectiveness. Ironically, the debate had come full circle and was resolved on exactly the same lines as the initial 1907 proposal put forth by Wilfrid Laurier, which the Government of India had rejected for ten years. The distinctive development, however, was that the national enabled a principle of pseudo-reciprocity and thus pseudo-universality to be inscribed within the passport system. The passport emerges here as a state document that purports to assign a national identity rather than a racial identity, and was a mechanism that would conceal race and the racist motivations for controlling mobility in the guise of a reciprocal arrangement between states described as national.

As I have noted in the introduction, much current scholarship on migration sees state control of migration as an inviolable and persistent feature of state sovereignty that dates not merely to Westphalia but to some even prior mythic time. In examining the formation of such control, however, one finds a rather different genealogy, one that points in the direction of the piecemeal development of state control over migration that eventually culminates in nation-state monopoly over migration. Colonial relations, moreover, were central to the emergence of such new formations that radically transformed understandings of the purview of sovereignty and thus were central to the formation of what is called sovereignty *doctrine*. Lest a certain presentism obscure the significance of these events and debates, it is important that we attend to the different context that framed these early twentieth-century discussions on migration. Thus, for instance, Frank Oliver's fear that the *Komagata Maru* incident was "an organized movement for the purpose of establishing as a principle the right that the people of India, and not the people of Canada, shall have the say as to who may be accepted as citizens of Canada" was a *serious* fear; for it was by no means certain that the matter would be resolved in favor of the barely emergent and fluctuating category of the "Canadian people." The solution offered by Gillian and Enthoven in India, which sutured sovereignty to migration control via the mechanism of nationality—simultaneously universal and particular—was an inspired and ad hoc solution; a solution, moreover, that had evaded all parties to the discussion for ten years. Viceroy Hardinge's appeal to "thoughtful people" and the "inherent rights" of states clarifies explicitly the historically contingent, eventful nature of the relation between national sovereignty and migration control. For the "inherent rights" Hardinge invoked had not made themselves apparent earlier to any party to the debate for a decade.

As with the debates attendant on state regulation of indentured Indians, what we can witness in these debates and correspondence is a critical remaking and transformation of the *norms* or *doctrine* of state sovereignty with regard to migration. But much had also changed from the early nineteenth-century regulations of indentured Indian migration that, we will recall, were, in part, justified in terms of protecting the sovereignty and liberty of the subject. The early twentieth-century regulations of "free" migration were justified solely in terms of protecting the sovereignty and liberty of the state. Moreover, the nineteenth-century regulations were instituted as an *exception* to the rule of "free" movement; the twentieth-century regulations, on the other hand, were offered as a *general* principle. Neither set of regulations, however, emerged without extensive contention, conflict, and debate that were related to the precise historical conditions that prevailed and could not simply utilize notions of sovereignty, be it of the subject or of the state, in uncontested terms.

The emergence of the nation-state as the first state formation to exercise a monopoly over migration indicates that control over mobility does not occur *after* the formation of the nation-state, but that the very development of the nation-state occurred, in part, to control mobility along the axis of the nation/race. In this way "nationality" comes to signify a privileged relation between people and literal territory. This relation, that sutures nationality to territory, is consolidated daily in the umpteen moments that passports are used. An attention to the resilience of such concrete technologies should also give us pause when we make pronouncements regarding the death of the nation-state and the emergence of an era of post-national globalization.[97]

Presently, we tend to think of the passport as not merely a necessary document for international mobility but as one that facilitates such movement. However, its history reveals that it marks its birth in an attempt to restrict movement along national lines that are explicitly racialized. Among the histories calcified in the history of the modern passport is a history of twentieth-century racism and a history of naturalizing the territorial boundedness of a national space as self-evidently the legitimate abode of certain people. The nineteenth-century interventions of the empire-state regulating the movement of indentured Indian labor were animated by a concern to distinguish that movement from the slave trade and justified by a rule of colonial difference anchored in the "civilizing mission." The twentieth-century state interventions regulating the movement of "free" Indians, embodied in the passport, were justified by a rule of colonial difference anchored in "nationality." The passport is thus a document that has effectively naturalized the rule of colonial difference in what one might call the "rule of postcolonial difference," where the marker of difference is not "race" but the putatively universal category of "nationality." It is therefore no surprise that today different "nationalities" have differential access to mobility.[98]

In History

A Colonial Genealogy of the Modern State

In British postcolonies the books of Enid Blyton were staple reading for young girls. As a consequence, growing up in India, like many others of my generation, I had avidly read her oeuvre.[1] Lacrosse was a popular sport in many a Blyton novel set in a girls' boarding school. In adulthood, I had systematically confused it with croquet and considered it a sport of European—most likely, French—provenance, played by young women in white dresses in much the manner it might be depicted in films by Merchant and Ivory—which is to say, in soft focus. Relatively recently, I learned that a genealogy of lacrosse would take one not to France and to Europe, but to the Iroquois/Haudenosounee of North America.

How, one might reasonably ask, might these seemingly random reminiscences be related to the subject of state control of migration and a colonial genealogy of the modern state? In July 2010, the Federation of International Lacrosse held its World Championship meet in Manchester, England. Since 1988, the Iroquois nation—which straddles Canada and the United States—has sent a team to compete in the event. The men's team, composed of members from both the United States and Canada, was unable to travel to the 2010 tournament since the United Kingdom refused to recognize the Haudenosounee passports of the team members and to issue the required visas.[2] Asserting their claim to sovereignty, the Haudenosounee have issued passports to their members since at least 1977. The objection of the UK functionaries oscillated between concerns, on the one hand, about whether the United States recognized the documents as entitling the bearers to entry back into the United States and, on the other, deeming the Haudenosounee passport document as technically substandard, lacking the necessary accouterments, such as security holograms, increasingly required in a post-9/11 world. Both Canada and the United States were asked to intervene, and then–U.S. Secretary of State, Hillary Clinton, using bureaucratic discretion, stepped in to grant a one-time exception and waiver to allow team members to use their Iroquois passports to guarantee reentry to the United States.[3] The United Kingdom, however, was intransigent.[4] With time running out and the tournament already under way, Clinton was also willing to issue team members emergency U.S. passports, a resolution unacceptable to the Iroquois team on two counts: first, it was unwilling, on principle, to thus cede sovereignty and accede to membership within the U.S. state.[5] Second, the team captain explained that such a resolution also posed a practical problem: championship rules required team members to register as national teams carrying the passports of their nation as documents of identity. As there was already a U.S. team participating in the tournament, if the Iroquois used U.S. passports, there would, effectively, have been two teams from the United States. Hence, while the U.S. passports would have enabled travel for the Iroquois, using them as documents of identity would have disqualified the team. The Native American provenance of the sport only added to the injustice and tragedy of the situation.[6]

I relate this episode not only because it encapsulates how the passport has become a necessary document for international mobility, or how it is the preeminent signifier of sovereignty and national identity, or how the ability to issue passports is itself now a marker of (contested) sovereign authority. I relate it also because it encapsulates—and invites an exploration of—a different set of colonial fissures at the heart of the modern state than those I have explored in this book. Such an exploration would reveal the less-than-settled status of settler states, including global hegemons like the United States, and enrich the colonial genealogy of the modern state I have presented here.

My analysis of colonial Indian migration shows how, in a period of less than a century, from the abolition of slavery in 1834, to the onset of World War I in 1914, state regulation of colonial Indian migration would undergo a series of radical shifts that, I have argued, are indicative of radical transformations in the state itself. In the wake of abolition, the "freedom" of the migrants was the preeminent concern of (quasi) state authorities. Articulated within an overarching paradigm that sought to facilitate the movement, there were two distinctive features of state regulation of Indian migration to the ex-slave plantation colonies: first, that it was instituted as an *exception* to the principle of free movement and, second, that migrants were required to consent to a labor contract that was held to verify that the movement was free. As I have shown in chapter 1, the rule of colonial difference animated and justified both these features, with the former representing a (temporary) alteration to understandings of state sovereignty with respect to mobility and the latter constituting a transformation in normative views of the free labor contract, wherein consent became of paramount concern. Though the centrality of consent to the free labor contract was the hallmark of nineteenth-century transformations to the labor contract, which became normalized in a range of locales over the course of the century, the rule of colonial difference that informed the temporary expansion of sovereignty, authorizing state intervention into free movement, retained its exceptional status. Indeed, even as indentured migration—free movement under a standardized state-authored and state-authorized contract—was extended to a range of sites, including those that were not erstwhile slave plantation economies, the exceptional status of such control remained.

This exceptional status, which contravened the principle of free movement, would, as we have seen in chapter 4, pose a grave problem to instituting more general, blanket legislative changes seeking to regulate—in order to prohibit—free movement to settler colonies in the early twentieth century. The primary, one can rightly say *only*, rationale of calls for such prohibition was racist and racial thinking, which glued the adjective "white" to a variety of "settler" colonies. In other words, "white-settler colonies" become such, in part, through their concerted efforts to regulate (predominantly Asian) migration and to become white. In detailing and analyzing the debates over Indian migration to Canada in the early twentieth century, I have demonstrated how the history of the modern passport, now the preeminent signifier of national and state identity, condenses how nationality, operating as an alibi for race, would become sufficiently significant as to aid the transformation of a world dominated by empire-states into a world dominated by nation-states. This transformation, born out of attempts to resolve the conundrum of how to prohibit migration conceived in racial terms, would result in an enduring remaking of state sovereignty and the inter-state system in specifically national terms.

Since our current academic discourse is preoccupied with discussions of sovereignty, the focus of my narratives in chapters 1 and 4 might have special appeal. But the Canadian state and many white Canadians were not the only ones to harbor racial anxieties regarding Indian migration. At the time, similar anxieties beset the newly formed South African state and many white South Africans. Unlike Canada, the route taken here to prohibit Indian migration did not center on notions of state or national sovereignty. As we have seen in chapter 3, the locus of concern was kinship relations, specifically polygamous marriage, which largely served as a surrogate for racial thinking even as it introduced additional elements into the equation. Moreover, in South Africa, national identity was most successfully activated by Indian protests in profoundly gendered terms, notably by Gandhi and elites participating in the 1913 *satyagraha* movement. Though the general view is inclined to view these events as iterations of anti-colonial—and thus, implicitly, anti-racist—struggles, they were, simultaneously, of a piece with racialized-national thinking.[7] Indeed, the trajectory of events was one where gendered tropes of national identity, tied to the politics of satyagraha, would eclipse other concerns of the Indian community. This turn of events did not succeed in mounting a radical challenge to the racist aims of the state that were, in fact, achieved in its requirement of marriage licenses combined with the cessation of indentured migration.

If, in chapters 3 and 4, we are exposed to the details of how two very different modes of establishing a racialized regime of migration were pursued at two different sites (namely, South Africa and Canada) and, in chapter 1, we learn how the abolition of slavery and the regulation of indentured Indian migration were intimately entwined—beyond merely noting the former as a cause for the latter—chapter 2 focused on the massive bureaucracy that took shape to manage migration. Over the course of a few short decades, we see the formation of an exceedingly complex bureaucracy to address the manifold aspects of indentured migration. Along with tracing the minutiae of the constant crises and the transcontinental, intra- and inter-imperial negotiations and correspondence that brought this bureaucracy into being, this chapter sought to make analytic sense of the plethora of rules and regulations that took shape. Deploying Foucault's notion of disciplinary power and attentive to important regularities across time and space, the chapter provided the lineaments of an elaborate migration regime while detailing the different regulations that emerged in different periods and with respect to different destinations. One advantage of this approach is that it enables us to understand both key aspects of the overarching bureaucracy as well as the modalities through which variations and specific rules and regulations were introduced. Another advantage is that it avoids either glossing over the variations (by making empirically inaccurate generalizations regarding indenture) or getting mired in specifics lacking an analytical frame (by understanding the regulations to a particular destination as entirely unique and not part of a broader system).

To my mind, it is of crucial significance that multiple aspects of the lineages of this bureaucracy, as also the lineages of the technologies traced in other chapters, continue to thrive in our present. As the debates and histories presented in the foregoing chapters vividly illustrate, these techniques and technologies did not develop autochthonously in spatial–territorial isolation. Rather, they developed as responses to historically specific conjunctures, through densely relational processes, and through practices of circulation and adaptation across a range of sites. An attention to such relations and circulations reveals that colonial formations were a critical aspect of the processes that produced the specific form of the modern state.

A Colonial Genealogy of the Modern State:
Epistemological Imperatives and Entailments

A primary aim of this book was to write a colonial genealogy of the modern state. As I conceived it, the burden of this endeavor was, equally, to wrench analysis of the state out of the tight grasp of a spatial–territorial closure, informed by methodological nationalism, as well as to avoid abstract meditations that reify the state. In my view, such approaches stymie an understanding of state regulation of migration and, in turn, what this can tell us about modern state formation. As important, they reinscribe the too-easy demarcation between the colonial state and the modern state, invariably falling prey to diffusionist and/or teleological understandings of state formation globally, which naturalize, rather than historicize, a "(neo)Europe/Other" framework. Such a perspective informs, for instance, the influential work of John Torpey on a history of the passport and state regulation of migration. Torpey is, I believe, correct to suggest that a vital, now-characteristic feature of the "statesness" of the modern state is a monopoly over migration control. But he errs in the view that the history of this outcome can be adequately told through a focus restricted to Europe. Working with a model of diffusion and imposition, Torpey believes his focus is warranted since "the dominance of Western states in the period examined [after the French Revolution] has been relatively clear cut, and the imposition of Western ways on most of the rest of the world has been one of the most remarkable features of the era."[8]

An assumption of a distinction—rather than a relation—between (neo) Europe and its Others, between the modern and the colonial, informs also important theorizations of the colonial state. As a result, in specifying the distinguishing features of the colonial state, it is, implicitly or explicitly, juxtaposed with the metropolitan state, the putative modern state. Such a juxtaposition and comparison is evident, for instance, in Partha Chatterjee's important and generative treatment (aspects of which have crucially informed my analysis). By Chatterjee's account, whereas the metropolitan/modern state would tend toward instituting an "impersonal, nonarbitrary system of rule of law," with the attenuation of legal and institutionalized differences premised on so-called ascriptive identities, the colonial state would tend toward an exacerbation of such identities (for example, of religion or of race).[9] For Chatterjee, due to the persistent mobilization of the rule of colonial difference in the strategies of rule, the modernizing promise of the colonial state was doomed to failure, embodying "the inherent impossibility of completing the project of the modern state without superseding the conditions of colonial rule."[10] Much like colonial thinking itself, Chatterjee here thus suggests a telos to state (trans)formation, wherein the metropolitan state emerges as the normative telos of the state, as such, and the colonial state emerges as its deviant, incomplete, shadowy double. By ascribing an aim, a hidden objective to the modern regime of power—whose *intended*, rather than *contingent*, outcome is to produce indistinguishable citizen-subjects—Chatterjee distinguishes metropolitan and colonial state formations as those where this objective is achieved in the former and thwarted in the latter.[11]

In a similar epistemological vein, George Steinmetz has recently offered two criteria that define the specificity of the colonial state, which are also framed within the normative horizon of the metropolitan state.[12] The first, what he calls the "*sovereignty* criterion," entails that "political sovereignty has been seized by a foreign political power."[13] The second, what Steinmetz, drawing on Chatterjee, calls "the *rule of difference* criterion," entails that "the indigenous population is treated by the conquering state as fundamentally inferior."[14] While Steinmetz's larger project is concerned with the *variability* of colonial state policies and wishes, particularly, to explain different "native policy" in the German Overseas Empire, he is also at pains to establish that colonial states *were* states, by ascertaining if they "meet" or "violate" Weberian criteria held to define states. Definitions here are granted a foundational status, true for all times and for all places, held to be abstractions that are not leaky or sullied by the empirical. Varied historical circumstances are thus assessed, or tested, against a set of criteria, understood as independent of context, with historical processes and events *measured* in terms of whether they adhere to or violate definitions and conceptual criteria.

As a consequence, the enormous historical variability Steinmetz discerns does not disturb, let alone transform, the definitional, conceptual criteria that define the state, as such. Even as Steinmetz's own additional criteria, which characterize the colonial state, threaten to undo Weber's definition and beg for an interrogation of how criteria are established, this is a path he eschews.[15] Instead, in a version of Chatterjee's colonial aberration thesis, Steinmetz concludes that if and when a state rejects the sovereignty and, especially, the rule of difference criteria, it "exits" from its colonial situation to become, simply, a normal, normative state.[16] Both Chatterjee and Steinmetz, as also others who study the colonial state, assume the territorial boundedness, demarcation, and distinction between the colonial state and the metropolitan state or, more broadly, of Europe and its Others. In fact, Steinmetz explicitly seeks to establish the territorial closure and relative autonomy of the colonial state, such that it can adhere to received understandings of Weberian criteria and be counted as a state, forgetting the historical occasion—and historical limitations—of Weber's formulations.

These approaches work, implicitly or explicitly, within a framework of comparison, contrasting metropolitan to colonial forms of state. The histories presented in the previous chapters point, unmistakably, to the lineaments of a world produced through processes of relationality and co-production, not autochthony. Coproduction, however, should not be understood as signaling the parity of state entities or entailing the equality of individual subjects; it can, as we have seen, signal the active and sustained production of a hierarchy of state entities and the inequality of subjects.[17]

Indeed, coproduction enables us to traverse the relational processes and conjunctural elements whose *outcome* is the material, institutional, ideological, and epistemological demarcations that characterize our world. While I have addressed this in more detail in the introduction, let me, once again, clarify my position, to make it less amenable to misinterpretation: I am emphatically *not* suggesting that there are no distinctions between the colonial state and the metropolitan state and that these did not have specific, if historically fluctuating, territorial reach. I *am* suggesting that modalities of analysis restricted to contrasting colonial state forms to metropolitan/modern state forms cannot grasp important aspects of the historical trajectories that produced the modern world and need supplementation by approaches more attuned to coproduction. With regard to migration, such perspectives enable us to recognize that territorial borders did not emerge first, only later and incidentally to be policed, monitored, guarded. Rather, borders are produced and experienced in specific forms and become porous or impassable with respect to certain subjects conceived in historically specific ways.[18] Indian migrants to the ex-slave plantation economies were largely conceived as ignorant subjects, in need of state protection to oversee and facilitate their movement; Indian migrants to white-settler colonies were conceived as illegitimate interlopers, who drove down wages and posed a threat to the social fabric of white, Christian societies. Though such were the overt justifications provided, their genuine rationales, as this book has revealed, were, respectively, to replace emancipated slaves with a "free," yet immobilized, labor force and to secure a white demography. Along the way of achieving these larger aims, a massive bureaucracy and a new nationalized subject, conceived as the national citizen, were also produced.

As a consequence, we have what are offered today as interrelated, axiomatic truths: that the national state has always held a monopoly over migration control and that the peoples of the world can be classified into those who are "citizens" and those who are "migrants," those who are "legal" and those who are "illegal," those who are properly "documented" and those who are "undocumented"; in sum, those who legitimately belong to a certain state-space and those who emphatically do not belong to it. The dominant logic that informs these classifications underlies the current global panic around the "problem" of migration, evident in both what we call the global North and the global South, which has yielded a panoply of state measures offered as "solutions." These include the ratification of extensive, seemingly constant fresh legislation across the world; the reorganization of state bureaucracies; the valiant efforts to systematize migration regimes (by seeking, for instance, to unambiguously distinguish forms of human movement into rigid classifications such as immigrants, refugees, trafficked victims, tourists, interlopers); the deployment of an ever-increasing "army" of patrol agents (who police the border in its multiple iterations, from airports and ports to "actual" international territorial divides); and the construction of insurmountable walls, thorny fences, and ever-larger detention centers. Alongside, there are non-state measures of intimidation undertaken by citizens, those held to legitimately belong to a state-space, ranging from the articulation of virulent anti-immigrant sentiments to eruptions of more immediate physical violence, which duplicate the state logic of exclusion and inclusion. Within the logic of these axiomatic truths, moreover, being classified as citizens or migrants, legal or illegal is not merely a politico–juridical matter; the categories have now also come to signify a subject's proximity and propensity to a venal criminality and are seen as indicative of deeper, underlying moral dispositions.

In contradistinction to this logic is the challenge posed today by activist organizations such as No One Is Illegal. The challenge, embodied in the very name of the group (which serves, simultaneously, as a slogan and a rallying cry), not only questions the particularities that inform the distinctions between legality and illegality, in an effort to eliminate, for instance, racist or sexist immigration regimes (a reformist agenda that characterizes numerous other, less radical, migrant rights groups), but, at its limit, questions the very notion that the modern subject is fully authorized via documentation by a certain state, that the modern subject is, first and foremost, the subject of a certain state, and that the state, in turn, has obligations only to those deemed its own legal, documented subjects (even as meeting these obligations remains persistently uneven and differential). Despite this radical challenge, the numerous chapters of No One Is Illegal largely direct their work toward the specific state in which they find themselves. Thus, even as No One Is Illegal articulates a radical critique, the critique itself is blunted in being forced to negotiate with the very entity whose authority is challenged: namely, the nation-state. In a peculiar way, then, this negotiation thus shores up, precisely, the authority of the national state as the central locus not only for the control and management of migration, but also for any transformation.

We thus arrive at the following conclusion. If the chief characteristic of colonial rule is a set of legal differentiations, which entail differential entitlements and differential treatment for different subjects, then today *all* states embody a *historically produced* colonial dimension, with the citizen/migrant distinction as a, perhaps *the*, primary axis of such differentiation. It is now well acknowledged that those conceived as migrants live—though in uneven ways—in the shadow of vulnerability and with the threat of illegitimacy, be they so-called international migrants or so-called internal migrants, like those who might have moved from Bihar to Mumbai. This distinction, between citizens and migrants, natives and foreigners, not only is now naturalized in non-liberal polities but also serves as the limit of entitlements and legitimacy in so-called liberal polities. Indeed, recently, several scholars have detailed the proliferation of these differentiated and differential legal regimes and their use in sites ranging from refugee camps and detention centers to more "ordinary" social and cultural exclusions across the globe.[19] Given such enduring legacies of colonial migrations, which are equally evident in the metropole as in the postcolony, it becomes important for us to robustly engage with what I have called a colonial genealogy of the modern state.

NOTES

Introduction

1. Well into the twentieth century, though the world was *dominated* by empire-states, they did not cover all possible state formations, admitting of a medley of other forms like kingdoms, city-states, fiefdoms, nation-states, among others. The empire-states, moreover, were a complex political form composed of a range of subpolities. Currently, the nation-state takes several different forms, including theocracies, liberal democracies, monarchies, and so on. For elaborations on the notion of the "empire-state," see Burbank and Cooper, *Empires in World History*; Mongia, "Race, Nationality, Mobility"; Mongia, "Interrogating Critiques of Methodological Nationalism." For another account that foregrounds the inescapable centrality of colonialism to the formation of a world composed of nation-states, see Kelly and Kaplan, *Represented Communities*.

2. Slavery in India was understood as qualitatively distinct from plantation slavery and was not legally abolished until 1843. For the complexities surrounding British understandings of, and responses to, slavery in India, see Chatterjee and Eaton, eds., *Slavery and South Asian History*; Major, *Slavery, Abolitionism and the Empire in India, 1772–1843*. See also the extensive report, *Letter from Government of India, February 1841; Report of Indian Law Coms., January 1841, on Slavery in E. Indies*, Parliamentary Papers (House of Commons) 28, no. 262 (1841).

3. For a detailed account of the sovereign and other powers vested in (and divested from) the East India Company, see Stern, *The Company-State*. By the early nineteenth century, the British Parliament had wrested away many of the company's earlier powers. On the related notion of "quasi sovereigns," see Grovogui, *Sovereigns, Quasi Sovereigns, and Africans*.

4. Secretary of State for the Colonies to Law Commissioners, India, May 25, 1836, quoted in Edward Lawford, Solicitor to the East India Company, to David Hill, June 12, 1838, *Papers Respecting the East India Labourers' Bill*, 2, IOR.

5. Statement showing the number of Coolies introduced into the Colony from Calcutta, from August 1, 1834, *Despatches from Sir W. Nicolay on Free Labour in Mauritius, and Introduction of Indian Labourers*, Parliamentary Papers (House of Commons) 37, no. 58 (1840), 41.

6. For details on the quantitative scale of these movements and an important corrective to the conventional wisdom that grossly underestimates Asian migration in the nineteenth and early twentieth centuries, see McKeown, "Global Migration, 1846–1940"; McKeown, *Melancholy Order*, 43–65. For a more recent overview, see Lucassen and Lucassen, eds., *Globalising Migration History*. For details on the distinction between the indenture system, which organized migration to the plantation economies that I examine here, and the *kangani* and *maistry* systems of migration from India to a variety of locales in South East Asia, Burma (Myanmar), and Ceylon (Sri Lanka), see Sandhu, *Indians in Malaya*; and Jain, *Racial Discrimination against Overseas Indians*.

7. On the lack of practical efforts in post-emancipation societies to ensure a transformation in the material condition of former slaves to the condition of freedom, see Holt, *The Problem of Freedom*; and Lowe, *The Intimacies of Four Continents*.

8. Recent attempts to address this oversight include Green and Weil, eds., *Citizenship and Those Who Leave*; and Moses, *Emigration and Political Development*. For analyses of the impact of Indian emigration on politics in India, see Sinha, "The Strange Death of an Imperial Ideal"; Sinha, "Whatever Happened to the Third British Empire? Empire, Nation, Redux"; and Charu Gupta, "'Innocent' Victims/'Guilty' Migrants."

9. Some of the significant work here includes Stuart Hall, "Notes on Deconstructing the 'Popular'"; Stuart Hall, "Cultural Identity and Diaspora"; Gilroy, *"There Ain't No Black in the Union Jack"*; Gilroy, *The Black Atlantic*; Clifford, "Traveling Cultures"; Clifford, *Routes: Travel and Translation in the Late Twentieth Century*; and Appadurai, *Modernity at Large*. For an important cultural studies exploration of "Indianness," which examines, in particular, its iterations in Trinidad, see Niranjana, *Mobilizing India*.

10. Some of the important work here includes Basch, Glick Schiller, and Szanton Blanc, *Nations Unbound*; Glick Schiller, "Transmigrants and Nation-States"; Portes, Guarnizo, and Landolt, "The Study of Transnationalism"; Pratt and Yeoh, "Transnational (Counter) Topographies"; and Levitt and Glick Schiller, "Transnational Perspectives on Migration."

11. On the problems with using "categories of practice" as "categories of analysis," see Brubaker, *Nationalism Reframed*. Relatedly, see also Cooper's discussion of "indigenous" and "analytical" categories in *Colonialism in Question*.

12. The centrality of state-space to definitions of transnational migration and transnationalism is evident, for instance, in Nina Glick Schiller's important work that is worth quoting at length. She writes: "I employ the word transnational to discuss political, economic, social, and cultural processes that extend beyond the borders of a particular state, include actors that are not states, but are shaped by the policies and institutional practices of states." Or, again: "Transnational migration is a pattern of migration in which persons, although they move across international borders and settle and establish social relations in a

new state, maintain social connections within the polity from which they originated. In transnational migration, persons literally live their lives across international borders. That is to say, they establish transnational social fields." See Glick Schiller, "Transmigrants and Nation-States," 96. Her definitions here are dependent on relatively stable understandings of the state, a position that cannot then account for transformations, including its "nationalization."

13. For a discussion of how migration regimes become a centralized, federal matter, with inter-state borders as the chief locus of concern, see McKeown, *Melancholy Order*.

14. Among others, Ralph Waldinger and David Fitzgerald have pointed to the anachronism embedded in the term *transnationalism* and noted that social identities might well not be organized in national terms. See Waldinger and Fitzgerald, "Transnationalism in Question." On the latter point, see also Markovits, *The Global World of Indian Merchants, 1750–1947*. Scholars of transnationalism are, of course, aware of the problems posed by this anachronism; their attempts to address it, however, have not been satisfying or consistent. For instance, in responding to Waldinger and Fitzgerald's charge of the anachronism embedded in transnationalism, Glick Schiller and Levitt point to the literature within transnational migration studies that documents how migrants, to the United States for instance, came to identify as "nationals" from their "nation-state of origin" as a result of the discrimination they faced. Left uninterrogated is *why* and *how* racial discrimination is resolved by way of, and sutured to, identity conceived in national terms: the issue explored by cultural studies scholars I have briefly addressed above. Also problematic is the formulation of "nation-states of origin" that assumes, rather than explains, states conceived in national terms. See Glick Schiller and Levitt, "Haven't We Heard This Somewhere Before?"

15. The issue is not substantively addressed in a range of landmark studies, including, for example, Anderson, *Imagined Communities*; Gellner, *Nations and Nationalism*; Chatterjee, *Nationalist Thought and the Colonial World*; Smith, *The Ethnic Origins of Nations*; Hobsbawm, *Nations and Nationalism Since 1780*.

16. Brubaker, *Nationalism Reframed*, 19.

17. Sewell, "Three Temporalities," 263.

18. See, for example, Salter, *Rights of Passage*; Green and Weil, eds., *Citizenship and Those Who Leave*; McKeown, *Melancholy Order*.

19. See, for example, Calavita, "U.S. Immigration and Policy Responses." It is worth pointing out that the 1882 U.S. Chinese Exclusion Act is a significant piece of this centralization of migration control.

20. Currently, barring *emigration* is seen as an index of a totalitarian state; barring immigration, common in liberal and nonliberal polities alike, is simply seen as the legitimate purview of the state. Quite the reverse understanding of the liberal state—as one that did not prohibit *immigration*, but might well prohibit *emigration*—prevailed in the nineteenth century.

21. Such exceptions include Torpey, *The Invention of the Passport*; McKeown,

Melancholy Order; and Mongia, "Historicizing State Sovereignty." Attributing the legitimacy of state control over mobility to the 1648 Treaty of Westphalia is vigorously assumed and advanced by, for instance, James Hollifield. See "The Emerging Migration State." It is reiterated by Alejandro Portes and Josh DeWind, who make it into a foundational, definitional matter and write: "*By definition*, states seek to regulate what takes place within their borders and what comes from outside." In this view, not only are borders natural, naturalized, self-evident, and static; investigations into the specificities of "what" precisely comes from "outside" and the variability of state forms are also rendered irrelevant, replaced by a continuist, untroubled history that traverses centuries. See Portes and DeWind, "A Cross-Atlantic Dialogue." Aristide Zolberg, via appeals to the Westphalian ideal, also takes this position, though, by my reading, his detailed work on U.S. migration policy and on innovations such as "remote control" (the varied practices of attempting to shape emigration in other states) points in the direction of, precisely, changes in sovereignty. See Zolberg, "Matters of State" and *A Nation by Design*.

22. Étienne Balibar's discussion of processes of "nationalization" is useful here. Arguing against teleological histories of the nation-state, in which a range of "qualitatively distinct events spread out over time, none of which implies any subsequent event" are interpellated and arranged as specifically prenational, Balibar suggests that we attend to how "*non-national* state apparatuses aiming at quite other (for example, dynastic) objectives have progressively produced the elements of the nation-state or . . . have been involuntarily 'nationalized' and have begun to nationalize society." See Balibar, "The Nation Form," 88.

23. For the moment, I will leave aside the issue of how, much like Weber's "ideal types," the Westphalian *ideal* does not have rigorous empirical coordinates, a matter that is relevant not only to how we understand this ideal's relation to colonized sites but also to formations in the heart of the metropole—or to colonizing sites. Indeed, a growing body of recent scholarship is concerned with debunking the "myth of Westphalia." See, for instance, Teschke, *The Myth of 1648*; Osiander, "Sovereignty, International Relations, and the Westphalian Myth"; Kayaoglu, "Westphalian Eurocentrism in International Relations Theory."

24. Abrams distinguishes between the state-system, "a palpable nexus of practice and institutional structure centered in government and more or less extensive, unified and dominant in any given society," and the state-idea, a set of notions "projected, purveyed and variously believed in in different societies at different times." In his view, to not reify the state, there are two approaches research might adopt. First, rather than study "the state" we might direct our attention to the specificities of the state-system and the state-idea. Second, studies might understand that "the state" is historically constructed and, thus, historicize it. My approach here combines elements of both approaches. See Abrams, "Notes on the Difficulty of Studying the State (1977)," 82.

25. I return to this issue at more length in the epilogue.

ture of colonial worlds, in terms of larger state structures such as sovereignty, see Benton, *A Search for Sovereignty*, and, in terms of domains such as intimate relations and everyday practices of rule, see Stoler, *Race and the Education of Desire*. See also Mawani, *Colonial Proximities*.

42. Chatterjee, *The Nation and Its Fragments*, 16–22. Giorgio Agamben notes that "necessity" is the first rationale for a state of exception. See Agamben, *State of Exception*.

43. For important treatments of the evolutionary temporal sequencing that attends a colonial logic, taking the varied forms of the civilizing mission, developmentalism, or modernization, see Fabian, *Time and the Other*, and Chakrabarty, *Provincializing Europe*. For an elaboration of developmentalist thinking in British liberal thought and the necessary tutelage natives required to be schooled into full political subjectivity, see Mehta, *Liberalism and Empire*.

44. Karuna Mantena distinguishes the two as universalist and culturalist forms of imperial ideology, respectively, with the latter gaining ground from the late nineteenth century. See Mantena, *Alibis of Empire*.

45. For discussions of an implicit or explicit comparative dimension in colonial thinking and the production of hierarchy, see Mongia, "Historicizing State Sovereignty"; Chatterjee, *The Black Hole of Empire*; and Mantena, *Alibis of Empire*.

46. For a textured account of this differential treatment, see Kolsky, "Codification and the Rule of Colonial Difference" and *Colonial Justice in British India*. See also Liu, "Legislating the Universal."

47. There is now a large and growing body of literature pursuing such research. Important work in this vein includes Burton, *Burdens of History*; Stoler, *Race and the Education of Desire*; Cooper and Stoler, eds., *Tensions of Empire*; Catherine Hall, *Civilizing Subjects*; Stoler, ed., *Haunted by Empire*; Hall and Rose, eds., *At Home with the Empire*; Ballentyne and Burton, *Empires and the Reach of the Global, 1870–1945*; and Lowe, *The Intimacies of Four Continents*.

48. See, for instance, Subrahmanyam, "Connected Histories"; Subrahmanyam, *Explorations in Connected History*; Hofmeyr, "The Black Atlantic Meets the Indian Ocean"; Hofmeyr, "Universalizing the Indian Ocean"; and Amrith, *Crossing the Bay of Bengal*.

49. For an extended elaboration of the production of political communities and the citizen/migrant distinction in the contemporary world, see Bridget Anderson, *Us and Them*.

50. Torpey, *The Invention of the Passport*.

51. I am aware that my formulation has evocations of Gyan Prakash's discussion of the "colonial genealogy of society." Prakash's discussion characterizes the colonial state as one premised on force and coercion and as distinct from the state as it developed in the metropolis. For Prakash, following Ranajit Guha, "the defining feature of the colonial state was its externality." While not abandoning the coercive aspects of the colonial state, by "a colonial genealogy of the modern

state" I wish to foreground the entanglements, rather than the distinctions, between the metropolitan/modern state and the colonial state. Such a perspective, as I will elaborate in the epilogue, enables us to grasp the colonial dimension inherent in *all* modern state formations. See Prakash, "The Colonial Genealogy of Society."

52. Foucault, "Nietzsche, Genealogy, History."

53. A key element of British rule was the legitimizing function of the "rule of law" that was consistently distinguished from the illegitimate, arbitrary rule thought to characterize "Oriental despotism." For a discussion of this duality, see Hussain, *The Jurisprudence of Emergency*, 35–68.

54. Anghie, *Imperialism, Sovereignty and the Making of International Law*.

55. The state's memory—or the official archive—is patchy and partial for a host of reasons, ranging from what was deemed worthy of archiving to the deterioration and loss of documents. For a discussion of how silences enter the archive, see Trouillot, *Silencing the Past*. For a meditation on the relationship between the archive and memory, see Derrida, *Archive Fever*.

56. Thus, in India, over time, the matter of migration came under the purview of the following different departments and branches (in turn, under constant reorganization): the Department of Home Affairs (until 1871); the Emigration Branch of the Department of Revenue, Agriculture and Commerce (from 1871 to 1879); the Emigration Branch of the Department of Home, Revenue, and Agriculture (1879–81); the Emigration Branch of the Department of Revenue and Agriculture (1881–1905); the Emigration Branch of the Department of Commerce and Industry (1905–20); the Emigration Branch of the Department of Commerce (1920–21); the Emigration Branch of the Department of Revenue and Agriculture (1921–23); the Overseas Branch (1923–32); the Lands and Overseas Branch (1932–38); the Overseas Section (1938–41) of the Department of Education, Health and Lands (which held the portfolio from 1923 to 1941); and the Indians Overseas Section of the Department of Commonwealth Relations (1944–49). In what we now know as "India," the Department of Commonwealth Relations was designated the Ministry of External Affairs on August 29, 1947, exactly two weeks after formal Indian independence.

57. Since most migrants were illiterate, their thumbprint served in lieu of a signature.

58. This usage was particularly common for indentees to Fiji. See Lal, *Girmitiyas*.

59. Exceptions to this trend include Hugh Tinker's classic study, *A New System of Slavery*, and, more recently, Rachel Sturman, "Indian Indentured Labor and the History of International Rights Regimes." My approach, however, differs from these studies.

60. This misunderstanding informs, for instance, John Comoroff's attempt to analyze and characterize the colonial state. While recognizing its analytical significance, he understands the "capillary" form of power (i.e., disciplinary power)

as displacing other forms of power. See Comoroff, "Reflections on the Colonial State, in South Africa and Elsewhere." For a critique of how sovereignty has become a "residual category" in much Foucauldian scholarship, see Singer and Weir, "Politics and Sovereign Power."

61. Chakrabarty, *Provincializing Europe*, 15.

62. On the impact of utilitarian ideas on British rule in India, see Stokes, *The English Utilitarians and India*. For a discussion of Foucault, the panopticon, and colonialism, see Martha Kaplan, "Panopticon in Poona."

63. McKeown, *Melancholy Order*, 44.

64. For instance, McKeown, *Melancholy Order*; and Lake and Reynolds, *Drawing the Global Color Line*.

65. This can be understood as part of what Mrinalini Sinha has called the "imperial-nationalizing" conjuncture that sought to rethink and remake empire as composed of different nationalities. Sinha, "Premonitions of the Past," 825.

66. While monogamous, heterosexual marriage offers a potential avenue for mobility, this is not to deny the complex—and differentially applied—scrutiny, policing, and evidentiary apparatus that surrounds the verification of marriage. For an account that points to such scrutiny, as also the means migrants use for subverting state mechanisms, see Kim, "Establishing Identity."

Chapter 1: The Migration of "Free" Labor

Epigraph: Despatch from Lord Stanley, Secretary of State for the Colonies, to Sir Lionel Smith, Governor of Mauritius, January 22, 1842, *Correspondence Relative to Indian Labor in Mauritius*, Parliamentary Papers (House of Commons) 30, no. 26 (1842): 31.

1. Prakash, "Colonialism, Capitalism, and the Discourse of Freedom," 10.

2. Liberal doctrine had garnered for itself a large portion of the credit for the abolition of slavery and it is generally held that liberal and religious humanitarian abolitionist thought and activity, primarily within Britain, were responsible for the abolition of slavery. Eric Williams, in his book *Capitalism and Slavery*, disputes this thesis, arguing, rather, that abolition owed more to economic imperatives than to humanitarian and philanthropic motives. For what Williams discerns as the ambivalences and hypocrisies of those involved in abolition, see, especially, chap. 11. Williams reserves substantial scorn for William Wilberforce, heralded as the chief spokesman for, and architect of, abolition. Williams's work is the object of substantial controversy. See, for instance, the essays in Solow and Engerman, eds., *British Capitalism and Caribbean Slavery*.

3. Andrew Sartori has recently noted that much scholarship on liberal political thought ignores one of its important dimensions, namely, political economy. I am sympathetic to his suggestion that a more robust history would "consider the long and complex relationship that liberal political thought has maintained with political economy." *Liberalism in Empire*, 22. Sartori seeks to produce a

history of "vernacular liberalism" by charting how the Lockean notion of what Sartori calls "the property-constituting capacity of labor" (7) not only became integral to liberalism in Britain but became a plausible claim of subsistence farmers in colonial Bengal. My aim here is to explore the colonial dimensions and determinations that made consent central to the free labor contract, a feature that would become hegemonic across a range of locales. See Sartori, *Liberalism in Empire*.

4. Important accounts of this transformation, which offer different analyses, include Horwitz, "The Historical Foundations of Modern Contract Law"; Horwitz, *The Transformation of American Law, 1780–1860*, especially chap. 6; Atiyah, *The Rise and Fall of Freedom of Contract*; and Gordley, *The Philosophical Origins of Modern Contract Doctrine*.

5. The nineteenth century not only witnessed a transformation in terms of the centrality of will, or consent, to contract. It also witnessed, as David Lieberman points out, an expansion in the number and kinds of relationships that came under the legal category of contract, most notably, for our purposes, that labor relations would increasingly move *out of* the domain of master and servant law and *into* the domain of contract law. This expansion led Henry Maine, in his 1861 treatise *Ancient Law*, to remark on "the largeness of the sphere [now] occupied by contract" in striking contrast to earlier times. See Lieberman, "Contract Before 'Freedom of Contract'"; Maine, quoted in Lieberman, 92. Despite the expansion of the sphere occupied by contract, Barry Wright notes that Indian indenture contracts more closely resembled master and servant regulations, particularly in their incorporation of employee criminalization for breach of contract, than they did free contracts. See Wright, "Macaulay's India Law Reforms and Labor in the British Empire." See also Michael Anderson, "India, 1858–1930."

6. Gordley, "Contract, Property, and the Will," 67.

7. The tendency to see the contracts made by ex-slaves, apprentices, or indentured labor as "deviations" from a stable norm prohibits considerations both of how the "norm" itself was in flux and of how the "deviations" might have been central to the changes that occurred in reconstituting the norm. In this way, the far-reaching, and often profoundly exploitative and coercive, implications of new laws are examined not as constitutive of changes in legal regimes, but simply as (stray) instances of exploitation that survived despite events such as abolition.

8. Hay and Craven, "Introduction," 31.

9. Studies that complicate this trend include Engerman, ed., *Terms of Labor*; Steinfeld, *Coercion, Contract, and Free Labor in the Nineteenth Century*; Cooper, Holt, and Scott, *Beyond Slavery*; Brass and van der Linden, eds., *Free and Unfree Labour*; Stanziani, "Local Bondage in Global Economies"; and Stanziani, *Bondage*.

10. Madhavi Kale suggests that historians have too readily accepted that slave emancipation led to a labor shortage due to decreased participation of ex-slaves in plantation labor. By her account, the need for more labor can be attributed to

the imperative to enhance sugar cultivation and production rather than to maintain it at the levels reached before emancipation. See Kale, *Fragments of Empire*, 56–65. For petitions from planters in Mauritius attempting to secure labor from India, see Petitions Relative to Immigration, 1829–1836, RC 24, NAM, and Petitions Relative to Immigration, 1837, RC 25, NAM. Later, planters would consider a host of locations, including Madagascar, Muscat, and Pondicherry, as sites to secure labor. See Letters Sent by Emigration Committee, 1839–1844, RD 57, NAM.

11. It is estimated that about 20,000 Indian slaves came to Mauritius prior to abolition. While Britain had abolished the slave trade in 1808 and France in 1817, the volume of illicit traffic was very large and continued for most of the nineteenth century. In fact, as Monica Schuler writes, "the antislave squadrons of Britain, the United States and France diverted an estimated 160,000 Africans from the slave trade between 1810 and 1864, less than ten percent of the estimated imports into the Americas between 1810 and 1870." See Schuler, "The Recruitment of African Indentured Labourers for European Colonies in the Nineteenth Century." See also Petitions Relative to Slavery, RC 23, NAM. In addition, Indian convicts were also transported to Mauritius, both prior to and after abolition. See Clare Anderson, *Convicts in the Indian Ocean*, and Letters Received from Calcutta, 1827–1835, RA 341, NAM.

12. Petition to the Queen from Planters, Traders, and Other Inhabitants of Mauritius, May 18, 1839, *Despatches from Sir W. Nicolay on Free Labour in Mauritius, and Introduction of Indian Labourers*, Parliamentary Papers (House of Commons) 37, no. 58 (1840): 7.

13. Statement showing the number of coolies introduced into the colony from Calcutta, from August 1, 1834, *Despatches from Sir W. Nicolay on Free Labour in Mauritius, and Introduction of Indian Labourers*, Parliamentary Papers (House of Commons) 37, no. 58 (1840): 41.

14. For the precise sequence of events attendant on John Gladstone arranging this departure, see Kale, *Fragments of Empire*, 13–37. See also Tinker, *A New System of Slavery*, 63–64.

15. Despatch from Governor Light to the Marquees of Normanby, September 5, 1839, *Reports or Despatches of Governor of British Guiana, Respecting Hill Coolies Introduced into the Colony*, Parliamentary Papers (House of Commons) 34, no. 77 (1840): 14. According to the remarkable diary of Theophilus Richmond, the surgeon on the *Hesperus*, one of the men had fallen overboard and drowned while the other had intentionally flung himself overboard and committed suicide, reportedly overtaken by grief that his father and brother had died of cholera and predeceased him. Richmond, *The First Crossing*, 92–94.

16. Kale, *Fragments of Empire*, 15.

17. *Report by Mr. Geoghegan on Coolie Emigration from India*, Parliamentary Papers (House of Commons) 47, no. 314 (1874): 2 (hereafter cited as Geoghegan's Report). On the sovereign characteristics of the British East India Company and other entities, see Stern, *The Company-State*. See also Benton, *A Search for Sov-*

ereignty, for a discussion of the multilayered nature of sovereignty in the early modern world.

18. Edward Lawford, Solicitor to the East India Company, to David Hill, June 12, 1838, *Papers Respecting the East India Labourers' Bill,* 3, IOR; Secretary to the Colonial Office/Supreme Government to Law Commissioners, India, May 25, 1836, quoted in Edward Lawford to David Hill, June 12, 1838, 2.

19. Response of Law Commissioners, quoted in Geoghegan's Report, 3. Thomas Babington Macaulay was the head of the Law Commission at the time.

20. Hollier Griffiths to G. F. Dick, Colonial Secretary, Mauritius, November 20, 1835, Enclosure in no. XIX, *Papers Respecting the East India Labourers' Bill,* 48, IOR.

21. Secretary to the Colonial Office to Law Commissioners, India, May 25, 1836, quoted in Edward Lawford, Solicitor to the East India Company, to David Hill, June 12, 1838, *Papers Respecting the East India Labourers' Bill,* 2, IOR.

22. Hollier Griffiths to G. F. Dick, Colonial Secretary, Mauritius, November 20, 1835, Enclosure in no. XIX, *Papers Respecting the East India Labourers' Bill,* 49, IOR.

23. Hollier Griffiths to G. F. Dick, 48–49.

24. Hollier Griffiths to G. F. Dick, 49.

25. Hollier Griffiths to G. F. Dick, 50 (emphasis added).

26. Hollier Griffiths to G. F. Dick, 50.

27. Hollier Griffiths to G. F. Dick, 50.

28. Plender, *International Migration Law,* 38–50.

29. For a discussion of some of these points, see Dummett, "The Transnational Migration of People Seen from a Natural Law Tradition."

30. Plender, *International Migration Law,* 43.

31. Locke, *The Second Treatise of Government,* 67–68.

32. Locke, *Second Treatise,* 69.

33. Translation of a Report by P. D'Epinay, Procureur-Général, Mauritius, January 5, 1836, Enclosure in no. XIX, *Papers Respecting the East India Labourers' Bill,* 58, IOR (hereafter cited as Report by P. D'Epinay). Originally in French, D'Epinay's report was crafted in response to the objections raised by Griffiths and took the opportunity of schooling the latter on the complex admixture of French civil law and British common law in Mauritius, which spoke to the fact that, until 1810, Mauritius was a French colony.

34. Report by P. D'Epinay, 56.

35. Report by P. D'Epinay, 56.

36. Report by P. D'Epinay, 58.

37. Report by P. D'Epinay, 59.

38. Report by P. D'Epinay, 59.

39. Despatch from Lord Glenelg, Secretary of State for the Colonies, to Governor Sir William Nicolay, May 25, 1836, *Papers Respecting the East India Labourers' Bill,* 69, IOR.

40. Lord Glenelg to Governor Sir William Nicolay, 69.

41. Kale, *Fragments of Empire*, 20.

42. Tinker, *New System*, 64–65.

43. Petition from the Inhabitants of Calcutta to Alexander Ross, President of the Council of India, July 10, 1838, *Letter from Secretary to Government of India to Committee on Exportation of Hill Coolies; Report of Committee and Evidence*, Parliamentary Papers (House of Commons) 16, no. 45 (1841): 148–49. For further details on the meeting, see Hossain, "Protests at the Colonial Capital."

44. Representation of the Merchants of Calcutta to Alexander Ross, President of the Council of India, July 31, 1838, *Letter from Secretary to Government of India to Committee on Exportation of Hill Coolies; Report of Committee and Evidence*, Parliamentary Papers (House of Commons) 16, no. 45 (1841): 147.

45. H. T. Prinsep, Secretary to the Government of Bengal, to Committee appointed to inquire into the abuses alleged to exist in the exportation of Hill Coolies, August 1, 1838, *Letter from Secretary to Government of India to Committee on Exportation of Hill Coolies; Report of Committee and Evidence*, Parliamentary Papers (House of Commons) 16, no. 45 (1841): 3. See also Tinker, *New System*, 64–65.

46. Tinker, *New System*, 66. See also Hossain, "Protests at the Colonial Capital."

47. Report of the Committee appointed to inquire into the abuses alleged to exist in exporting from Bengal Hill Coolies and Indian Labourers, *Letter from Secretary to Government of India to Committee on Exportation of Hill Coolies; Report of Committee and Evidence*, Parliamentary Papers (House of Commons) 16, no. 45 (1841): 4 (hereafter cited as Dickens Committee Report).

48. If one consults the archives in India it appears that Major Archer vanished without any comments on the matter. This is the view of Mr. Geoghegan, who, in 1871, undertook to prepare what would become a definitive report on the history and functioning of the indentured labor system. This is also the view of later historians such as Hugh Tinker and Madhavi Kale. However, though Archer departed from India and left the committee, he remained an active participant in the debate. Archer wrote to Lord Russell, supporting the resumption of indentured migration premised on the "indolence" of the slaves. See M. D. North-Coombes, "From Slavery to Indenture." In addition, Archer also submitted a long letter to the M.P. and antislavery activist S. Lushington. This letter was published with a sensationalist title: "Free Labour versus Slave Labour: A Letter to the Right Hon. S. Lushington, M.P. and the Opponents of Free Labour, Shewing that in their Opposition to Emigration from India to the British Colonies they are Virtually Encouraging the Slave Trade." My thanks to Riyad Koya for sharing the latter document with me.

49. Dickens Committee Report, 5.

50. Dickens Committee Report, 5.

51. Dickens Committee Report, 9.

52. Minute by Mr. Dowson, October 16, 1840, *Letter from Secretary to Government of India to Committee on Exportation of Hill Coolies; Report of Committee and Evidence*, Parliamentary Papers (House of Commons) 16, no. 45 (1841): 13–14 (hereafter cited as Minute by Mr. Dowson).

53. Minute by Mr. Dowson, 13.

54. Minute by Mr. Dowson, 14.

55. Dickens Committee Report, 6.

56. Minute by Mr. Dowson, 15.

57. Bhabha, "The Other Question."

58. Dickens Committee Report, 9.

59. Madhavi Kale's study, which attempts to move away from an analysis concerned with ascertaining the freedom of the migrants, is a rare exception to this rule.

60. Tinker, *New System*, xiv.

61. Northrup, *Indentured Labor in the Age of Imperialism, 1834–1922*, x.

62. For a discussion on the necessity for historicizing discourses on freedom, see Gyan Prakash's study of "bonded labor" in Bihar. Prakash, *Bonded Histories*. For an analysis that also seeks to move away from fixed, transhistorical understandings of freedom, but nonetheless retains a self-interested individual and the centrality of contract, see Steinfeld, *Coercion, Contract, and Free Labor in the Nineteenth Century*. For a critique of how contract is held to guarantee and embody freedom, see Marx, *Capital*, vol. 1, 76–88, 164–80; and Marx, *Grundrisse*, 156–73.

63. In general, there are three positions historians have adopted. First, those such as Brij Lal or Marina Carter, who hold that, in all likelihood, the circumstances and positions of different emigrants varied. Second, those such as Tinker and, more recently, Sugata Bose, who hold that the conditions did not facilitate consent. And, finally, those such as P. C. Emmer and, more recently, Ashutosh Kumar, who hold that migrants had sufficient knowledge of the circumstances and moved to escape immiserating conditions. Thomas Metcalf articulates perhaps a fourth position, suggesting that the debate "cannot be resolved on the basis of available documentation." Metcalf, *Imperial Connections*, 137. See Lal, *Girmitiyas*; Carter, *Servants, Sirdars, and Settlers*; Tinker, *New System*; Bose, *A Hundred Horizons*; Emmer, "The Meek Hindu"; Emmer, "The Great Escape"; Kumar, "*Naukari*, Networks, and Knowledge"; and Kumar, *Coolies of the Empire*.

64. "Questions to the Indian Labourers," Report from C. M. Campbell, T. Hugon, G. Villiers Forbes, and W. Bury to G. F. Dick, Colonial Secretary of Mauritius, November 13, 1838, *Despatches from Sir W. Nicolay on Free Labour in Mauritius, and Introduction of Indian Labourers*, Parliamentary Papers (House of Commons) 37, no. 58 (1840): 22.

65. "Questions to the Indian Labourers, Pamplemousses District," Report from E. Kelly and P. A. Heyliger to G. F. Dick, Colonial Secretary of Mauritius, September 7, 1839, *Correspondence Relative to the Introduction of Indian Labour-*

ers into Mauritius, *Report of Coms. of Inquiry*, Parliamentary Papers (House of Commons) 37, no. 331 (1840): 13.

66. See "Answers of the Indian Labourers," in Report from E. Kelly and P. A. Heyliger to G. F. Dick, Colonial Secretary of Mauritius, September 7, 1839, *Correspondence Relative to the Introduction of Indian Labourers into Mauritius, Report of Coms. of Inquiry*, Parliamentary Papers (House of Commons) 37, no. 331 (1840): 15–33, and "Answers of the Indian Labourers," in Report from C. M. Campbell, T. Hugon, G. Villiers Forbes, and W. Bury to G. F. Dick, Colonial Secretary of Mauritius, November 13, 1838, *Despatches from Sir W. Nicolay on Free Labour in Mauritius, and Introduction of Indian Labourers*, Parliamentary Papers (House of Commons) 37, no. 58 (1840): 18–34.

67. "Abstract of the Inquiry in Port Louis," Report from C. M. Campbell, T. Hugon, G. Villiers Forbes, and W. Bury to G. F. Dick, Colonial Secretary of Mauritius, November 13, 1838, *Despatches from Sir W. Nicolay on Free Labour in Mauritius, and Introduction of Indian Labourers*, Parliamentary Papers (House of Commons) 37, no. 58 (1840): 19–21.

68. C. Anderson, Special Magistrate, to G. F. Dick, Colonial Secretary of Mauritius, November 19, 1838, *Despatches from Sir W. Nicolay on Free Labour in Mauritius, and Introduction of Indian Labourers*, Parliamentary Papers (House of Commons) 37, no. 58 (1840): 35. For the sequence of events that brought Anderson's dissension to light, see Mongia, "Impartial Regimes of Truth."

69. C. Anderson to G. F. Dick, 36 (emphasis added).

70. This section draws extensively on Gordley, *Philosophical Origins of Modern Contract Doctrine*.

71. Gordley, *Philosophical Origins of Modern Contract Doctrine*, 163.

72. Gordley, *Philosophical Origins of Modern Contract Doctrine*, 167. See also Atiyah, *An Introduction to the Law of Contract*.

73. Gordley, *Philosophical Origins of Modern Contract Doctrine*, 208.

74. Gordley, *Philosophical Origins of Modern Contract Doctrine*, 162.

75. See Atiyah, *The Rise and Fall of Freedom of Contract*; von Mehren and Gordley, *The Civil Law System*. (Gordley has since revised his position.) For critiques of Atiyah, see Baker, "Review of *The Rise and Fall of Freedom of Contract*"; Lieberman, "Contract Before 'Freedom of Contract'"; and Gordley, "Contract, Property, and the Will."

76. This argument is advanced by Morton Horwitz, "The Historical Foundations of Modern Contract Law." See also Horwitz, *The Transformation of American Law, 1780–1860*, vol. 1, especially chap. 6. For an extended critique of Horwitz, including how he might have misread key cases, see Simpson, "The Horwitz Thesis and the History of Contract."

77. By Gordley's account, while there was a *doctrinal* change in how contract was understood in the work of legal treatise writers (with treatise writing—or the systematic mapping of doctrine and theories—itself a new, predominantly nineteenth-century development in the common law tradition), such doctrinal

changes seemed to have little, if any, impact on case law and the decisions of judges in Europe and the United States at the time. Thus, for Gordley, the doctrinal changes cannot be explained as a response to social and economic transformations, even though they would later come to impact legal understandings of contract. Gordley is especially disdainful of the proposition that the will theories bear a relationship either to ideological shifts or to more philosophical tracts in the nineteenth century.

78. For a superb analysis of the ravages of choice, framed within an understanding of freedom as equivalent to consent, in American Reconstruction, see Hartman, *Scenes of Subjection*, especially chap. 5.

79. Elizabeth Kolsky makes a similar point regarding the narrow national framework organizing legal history, writing that "empire is a framework that has eluded the notice of most legal historians." See Kolsky, "Codification and the Rule of Colonial Difference," 632. See also her *Colonial Justice in British India*. For a finely grained analysis that demonstrates how the selective incorporation of English master and servant law informed the making of a coercive, specifically colonial, labor regime in late eighteenth-century Madras, see Ahuja, "The Origins of Colonial Labor Policy in Late Eighteenth-Century Madras."

80. In his impressive overview of labor/employment law in colonial India, Michael Anderson suggests that, unlike in Britain and the United States, in India, "contractual fundamentalism," where the simple consent of the parties guaranteed the validity of contracts, did not grow deep roots. I am suggesting here that slavery abolition was a crucial aspect of the *rise* of such "contractual fundamentalism" that we can witness in the contractual relations that governed Indian indenture. See Michael Anderson, "India, 1858–1930," especially 444–45. It is important to note that with the growth of unions and organizations such as the International Labor Organization, such contractual fundamentalism was tempered.

81. The subject of their wide-ranging collection *Masters, Servants, and Magistrates in Britain and the Empire, 1562–1955* is the law of masters and servants across the British Empire. In their magnificent introduction they show how variegated iterations of this law, despite organizing the lives of most, if not all, subjects of empire (in Britain and without), fall within the purview of what they call "low law": though nested in statute, it was administered and enforced, often without record, by inferior magistrates and lay justices of the peace and, they write, has escaped the attention of the field of comparative law. As the essays collected in their volume show, to have an understanding of how legal regimes take shape necessitates an engagement with the domain of "low law." Moreover, departing from views that see a transition from pluralistic and multijurisdictional legal regimes to a state-centric legal regime from after the mid-nineteenth century (e.g., Benton, *Law and Colonial Cultures*), Hay and Craven suggest that master and servant law shaped and was integrated into state law of colonial legal regimes at a much earlier date. See Hay and Craven, "Introduction," especially 54–58.

82. For a superb analysis of the configurations of such other domains of contract law in late colonial India, see Birla, *Stages of Capital*.

83. Macpherson, *The Political Theory of Possessive Individualism*, 263–64.

84. Macpherson, *Possessive Individualism*, 48.

85. Hobbes, *Leviathan*, quoted in Macpherson, *Possessive Individualism*, 96.

86. Macpherson, *Possessive Individualism*, 96–98.

87. Marx, *Capital*, vol. 1, 165.

88. Macpherson, *Possessive Individualism*, 48.

89. Marx, *Capital*, vol. 1, 166.

90. Marx, *Capital*, vol. 1, 172.

91. Dickens Committee Report, 9.

92. Dickens Committee Report, 7–8.

93. Dickens Committee Report, 8.

94. Dickens Committee Report, 8–9.

95. Dickens Committee Report, 9.

96. It was Lord Stanley, when he had earlier also briefly been Secretary of State for War and the Colonies, from April 1833 to June 1834, who moved the Abolition Bill in Parliament and secured its passage. For details, see Holt, *The Problem of Freedom*, especially chap. 1.

97. The Emigration Committee, a planter organization in Mauritius, arranged to pay Charles Anderson £1500 to work on their behalf in London to get emigration from India approved. If he were successful, they would pay him an additional £500, for a total of £2000. See Emigration Committee to Charles Anderson, January 14, 1840, Letters Sent by Emigration Committee, 1839–44, RD 57, NAM.

98. C. Anderson, Special Magistrate, to G. F. Dick, Colonial Secretary of Mauritius, November 19, 1838, *Despatches from Sir W. Nicolay on Free Labour in Mauritius, and Introduction of Indian Labourers*, Parliamentary Papers (House of Commons) 37, no. 58 (1840): 35; Sir William Nicolay, Governor of Mauritius, to Lord John Russell, January 11, 1840, *Correspondence Relative to the Introduction of Indian Labourers into Mauritius, Report of Coms. of Inquiry*, Parliamentary Papers (House of Commons) 37, no. 331 (1840): 3.

99. C. M. Campbell, T. Hugon, G. Villiers Forbes, and W. Bury to G. F. Dick, Colonial Secretary of Mauritius, November 13, 1838, *Despatches from Sir W. Nicolay on Free Labour in Mauritius, and Introduction of Indian Labourers*, Parliamentary Papers (House of Commons) 37, no. 58 (1840): 18.

100. Reply of the Free Labour Association to a Report made by Messrs. T. Dickens, James Charles, and Russomy [*sic*] Dutt, Three Members of the Committee of Six, Instituted in Calcutta to Inquire into a Report on the Abuses Alleged to Exist in the Emigration of Indian Labourers to the British Colonies, June 4, 1841, *Correspondence Relative to Indian Labour in Mauritius*, Parliamentary Papers (House of Commons) 30, no. 26 (1842): 22.

101. Minute by Mr. Dowson, 13–14.

102. J. P. Grant's Minute on the Abuses Alleged to Exit in the Export of Coo-lies, March 1, 1841, *Papers on Exportation of Hill Coolies, from Government of India*, Parliamentary Papers (House of Commons) 16, no. 427 (1841), footnote (a), 30 (hereafter cited as Minute by J. P. Grant).

103. Minute by J. P. Grant, footnote (a), 31.

104. For descriptions and discussion on the content or material conditions of contracts attendant on abolition, including how they were articulated to vagrancy laws, incarceration and penal punishments, corporal punishments (mostly whipping), withholding and deduction of wages, requiring absence from work to be compensated by work of multiple times the length of absence, specific performance (that came to be called "industrial residence" with regard to In-dian migration), etc., see, for instance, Allen, *Slaves, Freedmen, and Indentured Laborers in Colonial Mauritius*; North-Coombes, "From Slavery to Indenture"; North-Coombes, *Studies in the Political Economy of Mauritius*; Trotman, *Crime in Trinidad*; Stanziani, *Bondage*, especially chap. 7; Turner, "The British Carib-bean, 1823–1838"; Mohapatra, "Assam and the West Indies, 1860–1920"; Chanock, "South Africa, 1841–1924"; Michael Anderson, "India, 1858–1930."

105. J. L. Austin's notion of the performative is now perhaps more associated with the work of Judith Butler than with its progenitor. But it is important here to briefly recapitulate Austin's formulation and explicate its importance to the relationship between contract, consent, and freedom. Austin coined the term *performative* to cover a certain species of speech acts, which, unlike other speech acts that Austin dubbed "constatives," did not simply describe or report events. Whereas the latter merely use language as a tool, or vehicle, to *convey* meaning, performatives affect a transformation in circumstances: they *do* things. Per-formatives thus demonstrate that language *does* things, and is far more than a mere vehicle. (Hence the title of Austin's text: *How to Do Things with Words*.) To use Butler's pithy formulation, performatives "bring into being that which they name." In explicating the notion of the performative, Austin's text brims with examples of promises, contracts, and covenants (including the promise/contract/covenant of marriage that, rightly, has received much attention from feminist and queer scholars). There is a close affinity and overlap between a promise, a contract, and a covenant. Indeed, much contract law derives from, or cites, a notion of the promise; a covenant, likewise, stands in close proximity to both. In Austin's schema, for a successful performative, a range of conditions must be met: chief among them is the "intention" of the parties, that they be "authorized" to engage in the performative in question, and that it be performed "sincerely," without fraud, duress, or mistakes (precisely some of the key ingredients of valid contracts). For Austin, "convention," an accepted procedure for doing things, and the "proper circumstances," are determining to the success or failure of a performative. In fact, Austin sees performatives as key elements of ritualistic or ceremonial acts. Derrida reworks Austin's notion to suggest that while intention and context remain important, they do not, by themselves, manage the scene

that produces successful performatives; rather, iterability is more germane. Moreover, in a classic deconstructionist move, Derrida suggests that the "failed" performatives, instead of being superfluous, are necessary to recognizing "successful" performatives. Working with and expanding both Austin and Derrida, Butler innovatively deploys the notion of the performative to stretch far beyond "speech acts." She suggests that "gender is a performative," produced through its endless iterations, intentional or otherwise, "failed" or "successful." I am suggesting here that the mere presence of a contract, a *ritual of consent*, brings into being that which it names: freedom. See Austin, *How to Do Things with Words*; Derrida, "Signature Event Context"; Butler, *Gender Trouble* and *Bodies That Matter.*

106. Dickens Committee Report, 9 (emphasis added).

107. Young's statement reads: "imperialism was not only a territorial and economic project but also a subject-constituting project." *White Mythologies*, 159.

108. Reply of the Free Labour Association to a Report made by Messrs. T. Dickens, James Charles, and Russomy [*sic*] Dutt, Three Members of the Committee of Six, Instituted in Calcutta to Inquire into a Report on the Abuses Alleged to Exist in the Emigration of Indian Labourers to the British Colonies, June 4, 1841, *Correspondence Relative to Indian Labour in Mauritius*, Parliamentary Papers (House of Commons) 30, no. 26 (1842), 28.

109. The ambivalence of colonial/racial discourse was at work vis-à-vis both Indian and African laborers. Thus, Dowson, for instance, accused African labor in Mauritius as primarily engaged in "the practice of keeping up the wages at a monopoly rate altogether ruinous," even as they were simultaneously held to have a "disposition to idleness." Concurring with Dowson, Major Archer, the elusive sixth member of the committee in India, had this to say: "It has been asserted that, the emancipated blacks would willingly work were they offered fair and adequate wages. This is a popular and very mischievous fallacy." Quoted in North-Coombes, "From Slavery to Indenture," 82.

110. The production of sugar in Mauritius rose dramatically from 1824–28, when it produced 17,119 metric tons, at the bottom of the rung of Britain's chief sugar-producing colonies, to 1854–58, when, with a production of 113,014 metric tons, it became Britain's most important sugar producer. See North-Coombes, "From Slavery to Indenture," 80.

111. Minute by J. P. Grant, footnote (a), 31.

112. Minute by J. P. Grant, 30.

113. C. Anderson to Lord Russell, May 1, 1840, *Correspondence Relative to the Introduction of Indian Labourers into Mauritius, Report of Coms. of Inquiry*, Parliamentary Papers (House of Commons) 37, no. 331 (1840): 194–95. Eric Williams has noted how invocations of impending "ruin" were a constant theme preceding the abolition of the slave trade and then of slavery. The theme continued in the post-abolition era. See Williams, *Capitalism and Slavery*.

114. C. Anderson to Lord Russell, May 1, 1840, 196.

115. C. Anderson to Lord Russell, May 1, 1840, 196.

116. C. Anderson to Lord Russell, May 1, 1840, 196.

117. Report of T. Hugon upon the Subject of Indian Emigration to Mauritius, July 29, 1839, *Correspondence Relative to the Introduction of Indian Labourers into Mauritius, Report of Coms. of Inquiry*, Parliamentary Papers (House of Commons) 37, no. 331 (1840): 184.

118. Minute by J. P. Grant, 31.

Chapter 2: Disciplinary Power and the Colonial State

1. J. P. Grant's Minute on the Abuses Alleged to Exit in the Export of Coolies, *Papers on Exportation of Hill Coolies, from Government of India*, Parliamentary Papers (House of Commons) 16, no. 427 (1841): 30 (hereafter cited as Minute by J. P. Grant).

2. For a discussion and critical assessment of the production of "impartiality" and its centrality to the regulation of Indian migration as well as to current state formations, see Mongia, "Impartial Regimes of Truth."

3. Minute by J. P. Grant.

4. This is especially important in our current moment, where efforts to un-ambiguously document identity, profile character, and ascertain consent are at the center of migration and other regimes. For recent critical explorations of procedures for documenting identity, including an interrogation of the premise that identity *precedes* its documentation, see Robertson, *The Passport in America*; McKeown, *Melancholy Order*; Caplan and Torpey, eds., *Documenting Individual Identity*; and Breckenridge, *Biometric State*. For an account of a range of identification procedures and practices (such as tattooing, photography, and fingerprinting—the last that found its birth in India) at work in colonial India, including documentation practices with regard to indentured migration, see Singha, "Settle, Mobilize, Verify." For an earlier, and important, elaboration of the use of, and difficulties with, deploying photography (entangled with physiognomy, phrenology, and statistics) as a means to produce a reliable representation of the body, particularly as it took shape around the documentation and identification of the criminal body and its enmeshment with practices of classification and the making of archival logics, see Sekula, "The Body and the Archive." Regrettably, given the contours of this project, I will not have occasion to substantively delve into issues of the documentation/production of individual identity.

5. Act XV of 1842, *Regulations and Orders by Government of Bengal for Protection of Coolies to and from Mauritius*, Parliamentary Papers (House of Commons) 35, no. 148 (1843): 1–7 (hereafter cited as Act XV of 1842).

6. Act XV of 1842, 6, 2, 1–7.

7. Act XV of 1842, 4.

8. Act XXI was passed on November 11, 1843. See *Report by Mr. Geoghegan on*

Coolie Emigration from India, Parliamentary Papers (House of Commons) 47, no. 314 (1874): 12–13 (hereafter cited as Geoghegan's Report).

9. Geoghegan's Report, 13.

10. Geoghegan's Report, 15.

11. Foucault, *Discipline and Punish*, 139.

12. Foucault, *Discipline and Punish*, 27.

13. For other accounts that explore bureaucracy and the rationalization of rule in colonial India, see Cohn, "The Census, Social Structure and Objectification in South Asia"; Bayly, *Empire and Information*; Richard Saumarez Smith, *Rule by Records*; and Raman, *Document Raj*.

14. In light of the discussion in the previous chapter, it is important to note here that the term *industrial residence* covered what in legal parlance is called "specific performance." The definition of "industrial residence" itself varied across colonies and over time.

15. Geoghegan's Report, 43. Geoghegan notes that such inconsistencies were of an "unaccountable nature" and "are not easy to understand," 43.

16. Geoghegan's Report, 41, 43.

17. Cover letter from Mr. Geoghegan to "A Note on Emigration from India," Agriculture, Revenue, and Commerce (Emigration), Proceedings No. 16–18 (A), NAI.

18. Now recognized as the father of comparative jurisprudence and of sociology of law, Maine was not deprived of due recognition in his day; in fact, he would go on from his post—and experience—in India to take up the newly established chair of historical and comparative jurisprudence at Oxford in 1869 and ended his career as Whewell Professor of International Law at Cambridge. Eric Stokes has examined the centrality of the "India experience" for important legal reformers and theorists, such as Maine and Fitzjames Stephen. See Stokes, *The English Utilitarians and India*.

19. See, for instance, Stokes, *The English Utilitarians and India*, 312–13. Important recent scholarship focuses attention on Maine's hostility to the applicability of utilitarian and "general" principles to colonial settings, seeing him, instead, as the architect of "indirect rule" and the necessity of "preserving custom," firmly committed to a slow, evolutionary process by which societies moved from "status to contract." See Mantena, *Alibis of Empire*; and Mamdani, *Define and Rule*.

20. Mr. Maine's "Statement of Objects and Reasons [For Act XIII of 1864]," quoted in Geoghegan's Report, 37 (hereafter cited as Mr. Maine's "Statement of Objects and Reasons [For Act XIII of 1864]"). The discrepancies between the terms of emigration for French and British colonies, and the more beneficial terms French planters had succeeded in securing, had been the source of substantial dissatisfaction for planters in British colonies.

21. Foucault, *Discipline and Punish*, 176.

22. Foucault, *Discipline and Punish*, 176–77.

23. Mr. Maine's "Statement of Objects and Reasons [For Act XIII of 1864]," 38.

24. Mr. Maine's "Statement of Objects and Reasons [For Act XIII of 1864]," 37.

25. Mr. Maine's "Statement of Objects and Reasons [For Act XIII of 1864]," 38.

26. Court of Directors to Governor General, May 3, 1843, Letters from the Court of Directors of the East India Company, Legislative Department, No. 8, 1843, NAI.

27. Court of Directors to Governor General, May 3, 1843. The letter suggests that in the event of the Governor General "authorizing the employment of the proposed officers in the interior of the country, it would be proper to require them to present any intended Emigrants before the Magistrate of the District in which they are engaged, or the nearest European functionary." Such communications, however, were frequently in the nature of suggestions and recommendations and thus did not have the "force of law."

28. Rules for the Guidance of the Protector, Medical Inspector of Emigrants, and the Colonial Emigration Agents at the Port of Calcutta, Under the Provisions of Act XIII of 1864, V/27/821/9, IOR (hereafter cited as Rules of 1864). See also Burbank to Bayley, October 15, 1863, Correspondence Regarding the Rules for the Guidance of all Persons Concerned in the Emigration of Native Labourers from Calcutta to the West Indies, Home Department (Public), December 31, 1863, Proceedings No. 77–78, NAI.

29. For an account that questions this view and seeks to provide an analysis of recruitment strategies with respect to Mauritius, see Carter, "Strategies of Labour Mobilisation in Colonial India." Recruitment and recruiters, who also often served in the capacity of headmen at plantations (and other sites) as agents of labor control and supervision, have received substantial attention in the scholarship on Indian indenture both to overseas colonies and to plantations in Assam. The scholarship has interrogated the position of Sirdars, or headmen, and returned migrants who served as recruiters, assessed the efforts of the Government of India to contain abuses, and engaged broader questions of the place of this specific labor agent and intermediary in economic transformations and labor history. See Metcalf, *Imperial Connections*, especially chap. 5; Roy, "Sardars, Jobbers, Kanganies"; Sen, "Commercial Recruiting and Informal Intermediation"; and Bates and Carter, "Sirdars as Intermediaries in Nineteenth-Century Indian Ocean Indentured Labour Migration." The close scrutiny of recruiters and recruitment practices was not unique to Indian indentured migration. It hounded discussions of, among others, Indian, Chinese, and Italian migration. For a discussion regarding Indian and Chinese migration and Italian migration, respectively, see McKeown, *Melancholy Order*, 87–88 and 118. McKeown claims that all migrant groups were accused of crimps and exploiters, "but few as regularly as Asians" (118).

30. Rules Relating to Colonial Emigration from the Port of Calcutta under Provisions of Act XXI of 1883, V/27/821/11, IOR (hereafter cited as Rules of 1883).

31. Rules of 1883.

32. Rules of 1864.

33. Radhika Singha notes a similar attempt at disassociation in terms of identification practices mobilized with regard to indentured labor. See Singha, "Settle, Mobilize, Verify," 164.

34. See, for instance, Kelly, *A Politics of Virtue*; Gillion, *Fiji's Indian Migrants*; Lal, "Kunti's Cry"; Niranjana, *Mobilizing India*; and Charu Gupta, "'Innocent' Victims/'Guilty' Migrants."

35. For an excellent description and analysis of the violence perpetrated by the overseers/managers and the sirdars on plantations in Fiji, see Lal, "Labouring Men and Nothing More," especially 139–41.

36. Noted in Gordon, Junior Secretary to the Government of Bengal, to Secretary to the Government of India, August 6, 1861, Communication from the Government of Bengal Submitting a Report of the Committee Appointed in October 1860 to Enquire into the System under which Emigration from This Presidency Is Conducted, Home Department (Public), December 7, 1861, Proceedings No. 10–12, NAI.

37. Gordon, Junior Secretary to the Government of Bengal, to Secretary to the Government of India, August 6, 1861.

38. Geoghegan's Report, 36. The cooks, Mr. Geoghegan notes, were appointed "after much protesting" by the Emigration Agents at Calcutta.

39. These chains of command are especially explicitly outlined in the "General Rules" section of the Rules accompanying Act XXI of 1883 and the Revised Edition issued in 1892. See Rules of 1883 and Rules Relating to Colonial Emigration under Provisions of Act XXI of 1883, Revised Edition, 1892, V/27/820/3, IOR (the latter hereafter cited as Rules of 1892).

40. Different patterns of migration control produce different sites as important. Eithne Luibhéid's important analysis demonstrates how the late nineteenth-century U.S. immigration control apparatus "served as a crucial site for the construction and regulation of sexual norms, identities, and behaviors." *Entry Denied*, x. Luibhéid effectively elaborates how the very form of migration control served, centrally, to aid the process by which new typologies of humans were generated and how the examination at the border was critical to this process. See Luibhéid, *Entry Denied*.

41. For an illuminating and provocative discussion concerning the policing of the route of laborers during a Hosay celebration procession in Trinidad, see Mohapatra, "The Hosay Massacre of 1884."

42. Mr. Maine's "Statement of Objects and Reasons [For Act XIII of 1864]," 38.

43. Due to climatic conditions and the length of the voyage, emigration was not permitted year round to all the colonies, except Mauritius. Thus, most depots, too, were operational for only part of the year.

44. Report from Surgeon G. B. Partridge, Government Medical Inspector of

Emigrants, to Captain Burbank, Officiating Protector of Emigrants, Calcutta, June 27, 1863, Enclosure in Home Department (Public), September 7, 1863, Proceedings No. 6–11, NAI (hereafter cited as Surgeon Partridge's Report).

45. Surgeon Partridge's Report.

46. Rules Relating to Emigration from the Port of Calcutta Under the Provisions of Act VII of 1871, V/27/821/10, IOR (emphasis in original; hereafter cited as Rules of 1871).

47. Rules of 1871. These regulations were also routinely violated, a point to which I will return. For an excellent and in-depth analysis of the rape and subsequent death of one passenger, Maharani, see Shepard, *Maharani's Misery*.

48. Theophilus Richmond, the surgeon on the *Hesperus*, the first ship to carry Indian labor to British Guiana in 1838, had the foresight to thus cordon off those suffering from cholera. In his diary, organized as a letter to his mother, he writes, "From the very commencement of its [the cholera] breaking out I had a place set apart, into which no one but myself and the sick were permitted to enter, and by thus cutting off all communication and using strong fumigations, I was so far successful in checking infection that on Thursday there were only 4 seizures and 2 deaths." Seven men had already died; more would follow. Richmond, *The First Crossing*, 90. By J. P. Grant's account, another man had died before the ship had even left Calcutta. See Minute by J. P. Grant, 17–18.

49. Correspondence on Rules Under Act VII of 1871, Agriculture, Revenue, and Commerce (Emigration, Part II), September 1872, Proceedings No. 66, NAI.

50. See, for instance, Tinker, *New System*; Kelly, *A Politics of Virtue*; Mohapatra, "'Restoring the Family'"; Brass and Bernstein, "Introduction"; Ramasamy, "Labour Control and Labour Resistance in the Plantations of Colonial Malaya"; Kelly, "Coolie as a Labour Commodity"; Lal, "Kunti's Cry"; Lal, "Understanding the Indian Indenture Experience"; Desai and Vahed, *Inside Indian Indenture*; Wahab, "In the Name of Reason"; Carter, *Voices from Indenture*; Reddock, "Freedom Denied"; Bahadur, *Coolie Woman*; Mahase, "'Plenty a dem run away.'" For a contemporaneous, firsthand account, see Sanadhya, *My Twenty-One Years in the Fiji Islands*. For an analysis that locates Sanadhya's text within a growing mass movement in India for the abolition of indenture, see Sinha, "Totaram Sanadhya's *Fiji Mein Mere Ekkis Varsh*." There is also, of course, a very vast and excellent literature concerning slave plantations. See, for instance, Sidney Mintz's classic, *Sweetness and Power*.

51. Rules of 1871.

52. Rules of 1864.

53. Rules of 1871.

54. Rules of 1883; Rules of 1892.

55. I thank Prabhu Mohapatra for the formulation of mapping the regulatory structure onto the knowledge structure.

56. Seeing indenture as part of "the expansion of modern forms of state

power," Rachel Sturman traces a connection between the regulatory regime of indenture and the emergence of international labor standards and a welfarist human rights regime. Sturman, "Indian Indentured Labor and the History of International Rights Regimes," 1442.

57. Correspondence on Rules Under Act VII of 1871, Agriculture, Revenue, and Commerce (Emigration, Part II) September 1872, Proceedings No. 66, NAI (emphasis in original).

58. Foucault, *Discipline and Punish*, 137.

59. This policy was explicitly clarified by the Governor General's Council in 1877: "We provide a medical inspection which is directed to testing the emigrant's fitness *for the voyage*, but we do not promise that he will be fit to labour when he reaches the journey's end; that is not our province: the colony must satisfy itself that it has an efficient agent in the country, and look to him, and to him only, for the supply of the men that it wants." Governor General's Council to Secretary of State, May 3, 1877, Home, Revenue, and Agriculture (Emigration), February 1880, Proceedings No. 4–29, NAI (emphasis in original). The policy was already in place in 1863; see Surgeon Partridge's Report.

60. Surgeon Partridge's Report.

61. Surgeon Partridge's Report (emphasis in original).

62. Surgeon Partridge's Report.

63. On the production of the modern body in the domain of colonial medicine, see Arnold, *Colonizing the Body*. For a discussion of the governmentalization of the colonial state, specifically with regard to sanitation and the colonial/modern body in India, see Prakash, "Body Politic in Colonial India," and *Another Reason*, especially chap. 5.

64. Ashutosh Kumar has examined the provisions for dietary requirements on board emigrant ships to assess what they tell us about how the colonial state accommodated the caste, religious, and regional dietary preferences of emigrants. See Kumar, "Feeding the *Girmitiya*."

65. Dr. Payne, Medical Superintendent, Emigration to Mauritius, to Thomas Caird, Emigration Agent for Mauritius, October 16, 1861, Home Department (Public), July 30, 1862, Proceedings No. 34–41, NAI.

66. Dr. F. J. Mouat, Inspector of Jails, Lower Provinces, to A. R. Young, Secretary to the Government of Bengal, July 31, 1858, Home Department (Public), October 1, 1858, Proceedings No. 25–29, NAI. Following the Indian revolt of 1857, Dr. Mouat headed the committee to ascertain if the Andaman Islands were a suitable site for a penal colony. On aspects of his explorations, see Clare Anderson, "Oscar Mallitte's Andaman Photographs, 1857-8."

67. Dr. F. J. Mouat to A. R. Young, July 31, 1858.

68. Dr. F. J. Mouat to A. R. Young, July 31, 1858.

69. For connections between indenture and regimes of incarceration and confinement, see Clare Anderson, "Convicts and Coolies."

70. Dr. Payne, Medical Superintendent, Emigration to Mauritius, to Thomas

Caird, Emigration Agent for Mauritius, October 16, 1861, Home Department (Public), July 30, 1862, Proceedings No. 34–41, NAI.

71. Instructions for the guidance of Surgeon Superintendents of Government Emigrant Ships regarding contagious fever, from J. M. Cuningham to J. Geoghegan, October 25, 1869, Home Department (Public), December 11, 1869, Proceedings No. 50–52, NAI.

72. "Proceedings of a Committee," from British Consular Agent, Pondicherry, to Protector of Emigrants, Madras, August 9, 1871, Correspondence on Rules Under Act VII of 1871, Agriculture, Revenue, and Commerce (Emigration Proceedings, Part II), September 1872, Proceedings No. 55, NAI.

73. Act XV of 1842.

74. Rules of 1864.

75. Rules of 1871.

76. Rules of 1883.

77. Rules of 1892.

78. Rules of 1892.

79. Foucault, *Discipline and Punish*, 189.

80. Foucault, *Discipline and Punish*, 189.

81. Rules of 1871.

82. Foucault, *Discipline and Punish*, 190.

83. Lal, "Labouring Men and Nothing More"; Behal, "Coolie Drivers or Benevolent Paternalists?"; and Amrith, "Indians Overseas?" have explored, with regard to Fiji, Assam, and Malaya, respectively, how there was no dearth of reports that indicated the acute violence of the plantation system, but that the system continued regardless of this information and in the interests of the plantocracy. Indeed, reports of abuse would sometimes cause the suspension of migration from certain districts or to certain colonies for a few years while inquiries were conducted, new rules established, and fresh guarantees secured. Complete cessation, however, was infrequent and usually implemented with regard to French and Dutch colonies and spoke more to inter-imperial rivalry than to greater abuse or infractions of the rules at these sites. On the latter point, see Stanziani, *Bondage*, chap. 7. See also Geoghegan's characterization of recruitment for the French colonies in Geoghegan's Report, 5.

84. Adopting such positions, we might include work such as Tinker, *New System*; Lal, "Kunti's Cry"; Beall, "Women Under Indenture in Colonial Natal"; Kelly, "Gaze and Grasp"; Carter, *Voices from Indenture*; Kale, *Fragments of Empire*; Shepherd, *Maharani's Misery*; and Mishra, "Indian Indentured Labourers in Mauritius."

85. Indeed, it was precisely the resort to this rationale of "good regulation" that had enabled the resumption of indenture in 1842.

86. Foucault, *Discipline and Punish*, 178.

87. As Francois Ewald points out, "Discipline is not necessarily normative." See Ewald, "Norms, Discipline, and the Law," 141.

88. Chris Biden, Emigration Agent, Madras, to Colonial Secretary, Mauritius, March 4, 1843, HA 83, NAM.

89. Geoghegan's Report, 45.

90. There are both important continuities and significant disjunctures between the bureaucratic regime of colonial Indian migration and the bureaucratic regime of poverty alleviation programs in postcolonial India that Akhil Gupta describes, even as the result of both is endemic violence. See Akhil Gupta, *Red Tape*.

91. My position and analysis here diverges from the view taken by Thomas Metcalf, who argues that the framework adopted by the government in India produced a progressively less exploitative system. Regarding, specifically, the process of recruitment for Natal and reports of abuse, he writes: "The periodic stoppages of indentured migration when complaints of abuse arose, as in Mauritius in 1839 and Natal in 1871, are evidence of a continuing concern [of the Government of India] for the welfare of migrants." Metcalf, *Imperial Connections*, 160–61.

92. Though there are intimate links and important analogies between bureaucratic exception/exemption and sovereign exception (as theorized by Giorgio Agamben), they require precise specification, and I would warn against generalized positions that conflate the two. On sovereign exception, see chapter 1, and Agamben, *State of Exception*.

93. Mongia, "Impartial Regimes of Truth."

94. See Adam McKeown, *Melancholy Order*, chap. 2; Sunil Amrith, *Migration and Diaspora in Modern Asia*; and Sunil Amrith, "South Indian Migration, 1800–1950."

95. See Claude Markovits, *The Global World of Indian Merchants*.

Chapter 3: Gendered Nationalism, the Racialized State, and the Making of Migration Law

1. For elaborations of this argument, see Tinker, *New System*, especially chap. 9; Reddock, *Women, Labor and Politics in Trinidad and Tobago*; Reddock, "Freedom Denied"; Kelly, *A Politics of Virtue*, especially chap. 2; Niranjana, *Mobilizing India*, especially chap. 2; Nijhawan, "Fallen Through the Nationalist and Feminist Grids of Analysis"; and Charu Gupta, "'Innocent' Victims/'Guilty' Migrants."

2. Tony Ballantyne, for instance, has argued for a refinement of the core-periphery model and for the "web-like" character of imperial formations that also admitted of "subimperial" centers. See Tony Ballantyne, "Rereading the Archive and Opening up the Nation State." For an important early explication of an "imperial social formation," see Mrinalini Sinha, *Colonial Masculinities*. For further elaboration, see Sinha, *Specters of Mother India*. See also Metcalf, *Imperial Connections*.

3. See my discussion of eventful approaches in the introduction as well as Brubaker, *Nationalism Reframed*; Sinha, *Specters of Mother India*, particularly "Introduction: The Anatomy of an Event," 1–22.

4. For an examination of a shift from a pluralistic and multi-centric legal order to a state-centric legal order, "in which the state has at least made, if not sustained, a claim to dominance over other legal authorities" (11), see Benton, *Law and Colonial Cultures*. See also Baxi, "'The State's Emissary.'"

5. This four-province structure remained in place until the post-apartheid era when South Africa was divided into ten administrative units.

6. For details on these factors, see Beall, "Women Under Indentured Labour in Colonial Natal, 1860–1911," 147; Pachai, *The International Aspects of the South African Indian Question, 1860–1971*, 6–7; Tinker, *New System*, 108.

7. Bhana and Brian, *Setting Down Roots*, 15.

8. For details on the circumstances and negotiations leading up to installing indentured migration to Natal, see Pachai, *International Aspects*, 2–7.

9. For an excellent analysis of the low female ratio of the regulation, including its explicit disinterest in the formation of a settler population and how it shaped an exploitative, gendered labor regime in Trinidad and British Guiana (with definite resonances with the socioeconomic formation of indentured Indians in South Africa), see Mohapatra, "'Restoring the Family.'"

10. Beall, "Women Under Indentured Labour," 148. This was unlike the situation in Trinidad and British Guiana where ex-indentured laborers had greater access to Crown and plantation lands. See Mohapartra, "Restoring the Family," 246–49.

11. Pachai, *International Aspects*, 8.

12. Pachai, *International Aspects*, 9.

13. Most ex-indentured laborers did not earn a wage sufficient to pay the tax. And, in fact, at the end of 1913 only 958 of the 10,805 liable to pay the tax did so. Bhana and Brian, *Setting Down Roots*, 61.

14. Historiographies of these events have largely adopted a Gandhi-centric focus and not engaged the issues in sufficient depth. For an analysis of historiographies of the 1913 satyagraha and their Gandhi-centric focus, see Mongia, "Gender and the Historiography of Gandhian *Satyagraha* in South Africa."

15. Swan, "Ideology in Organized Indian Politics, 1891–1948," 197, and Swan, *Gandhi*, 64.

16. The cessation was actually masterminded and managed through a complex series of confidential deals struck between the Union government of South Africa (without consultation with the concerned parties in Natal) and the Secretary of State for the Colonies in London. Huttenback, *Gandhi in South Africa*, chap. 8, and Telegram from Secretary of State for the Colonies, London, to Governor General, South Africa, January 4, 1911, *Union of South Africa, Correspondence Respecting a Bill to Regulate Immigration into the Union of South Africa*,

With Special Reference to Asiatics, Parliamentary Papers (House of Commons) 52, Cd. 5579 (1911).

17. For biographical information on Indian traders, see Swan, *Gandhi*, 2–10.

18. Until 1897, "every man over the age of 21 who either owned immovable property to the value of £50 or rented property to the value of £10 per annum was entitled to the parliamentary franchise." In 1897, after Natal had secured responsible government, the qualification was changed to deny the franchise to those who came from countries that did not already have representative institutions in place. Importantly, the number of ex-indentured Indians who would have the potential to vote was also on the rise. See Pachai, *International Aspects*, 2, 11.

19. On the emergence and circulation of a literacy test as a surrogate for racial thinking, see Lake, "From Mississippi to Melbourne via Natal"; and Lake and Reynolds, *Drawing the Global Color Line*.

20. Chanock, *The Making of South African Legal Culture, 1902–1936*, 29. In many jurisdictions, administrative law now forms a full-fledged branch of law that, in part, considers who can review discretionary decisions and when, since the exercise of such discretion is often in conflict with the notion of "due process."

21. Both the Cape and the Transvaal had toyed with the idea of using Indian indentured labor, but had decided against it. The Transvaal had, instead, recruited Chinese indentured labor on fixed three-year contracts between 1904 and 1907. For details on Chinese indentured labor in the Transvaal, see Richardson, "Coolies, Peasants, and Proletarians" and "Chinese Indentured Labour in the Transvaal Gold Mining Industry, 1904–1910." For a comparative assessment of Indian and Chinese indentured migrations to South Africa, see Harris, "Sugar and Gold."

22. See Gandhi, *Satyagraha in South Africa*, 25.

23. Huttenback, *Gandhi in South Africa*, 100.

24. Huttenback, *Gandhi in South Africa*, 100.

25. Bhana and Brian, *Setting Down Roots*, 194.

26. Bhana and Brian, *Setting Down Roots*, 194.

27. According to Gandhi, "[there was] less race hatred in the Cape Colony than in the other parts [of South Africa]" and there was a large proportion of Europeans in the Cape who "had kindly feelings . . . [and] a warm corner in their hearts for Indians." *Satyagraha in South Africa*, 37. Huttenback, *Gandhi in South Africa*, 98–99. See also Bhana and Brian, *Setting Down Roots*, 139–42.

28. This was the first condition laid down in 1843, when the British annexed Natal and laid the framework for the structures governing the Republic of Natalia. See Pachai, *International Aspects*, 2. This was also the substance of the Queen's Proclamation of 1858 when India came under the direct rule of the British Crown following the Rebellion of 1857.

29. The so-called Indian population was estimated at 15,000 in 1899, at the

outbreak of the South African war, and 10,048 in 1911 after the formation of the Union. Bhana and Brian, *Setting Down Roots*, 78, 194. Robert Huttenback puts the number at 17,000 at the outbreak of the war. *Gandhi in South Africa*, 103n.

30. These "locations" were later called "bazaars." Significantly, "locations" and "reserves" were already in place for the "native" population, particularly in Natal, and are the clear precursor to the full-fledged system of apartheid. Vis-à-vis the "native" population the segregation was carried out via the mechanism of "retribalization," the policy first put in place in Natal (a British and not a Boer territory) by Theophilus Shepstone, who was first Diplomatic Agent and then Secretary for Native Affairs.

31. For the complications surrounding the Transvaal as an independent republic but under the ultimate suzerainty of Britain and its relation to the passage of Law 3 of 1885, see Huttenback, *Gandhi in South Africa*, 102–9.

32. For correspondence regarding the recoding of racial exclusion as so-called non-racial "sanitary" exclusion as well as Indian opposition, see *Papers Relating to Grievances of Her Majesty's Indian Subjects in the South African Republic*, Parliamentary Papers (House of Commons) 71, C. 7911 (1895).

33. Lord Selborne, Governor of Transvaal, to Secretary of State for the Colonies, May 21, 1906. Department of Commerce and Industry (Emigration Proceedings—Part A), October 1906, Proceedings No. 3, NAI. The British government's objection to the policies against Indians in the Transvaal was more about asserting British authority over the affairs of the Transvaal than about protecting the rights of British Indians. For these details and the Transvaal government's delay in implementing the law, see Swan, *Gandhi*, 80–89, and Huttenback, *Gandhi in South Africa*, 102–22.

34. Indeed, the 1906 crisis generated the very term *satyagraha* for what had hitherto been called passive resistance. For details on coining the term *satyagraha*, see Gandhi, *Satyagraha in South Africa*, 109–10.

35. It is now seemingly acceptable practice to provide thumb- and fingerprints for a range of identification procedures. However, in 1906, fingerprints were required, by law, only from those deemed criminals and thumbprints were used primarily in lieu of a signature. From the point of view of the literate elite, the requirements for finger- and thumbprints were both an affront: the former since it coded them as criminals and the latter since it denied the fact of their literacy. For objections to requiring fingerprints, see Gandhi, *Satyagraha in South Africa*, 100–101. For detailed and illuminating analyses of Gandhi's ambivalence to thumb- and fingerprints in South Africa (and to state bureaucracies), see Breckenridge, "Gandhi's Progressive Disillusionment" and *Biometric State*, especially chap. 3.

36. As Gandhi recounts it, at one of the meetings held to discuss the ordinance one of the men "said in a fit of passion: 'If any one [*sic*] came forward to demand a certificate from my wife, I would shoot him on that spot and take the consequences.'" And Seth Haji Habib, one of the members of the deputation to

the Colonial Secretary, is reputed to have told him: "'I cannot possibly refrain myself if any officer comes and proceeds to take my wife's finger prints. I will kill him there and then and die myself.'" Gandhi, *Satyagraha in South Africa*, 101, 108. (I return to this point later.) Radhika Singha finds similar processes at work in India with regard to the identification procedures regarding female subjects. See her "Colonial Law and Infrastructural Power." For the unfoldings in South Africa, see Breckenridge, *Biometric State*, especially chap. 3.

37. The Chinese of the Transvaal also supported and were active participants in the early phases of the satyagraha movement. See Harris, "Gandhi, the Chinese and Passive Resistance." As the aphorism, attributed to Nelson Mandela, goes, "You gave us a lawyer, we gave you back a Mahatma."

38. For an analysis of how the regulation of migration moved from the domain of local authorities into the domain of centralized, federal authority in the twentieth century, see McKeown, *Melancholy Order*.

39. Mamdani, *Define and Rule*, especially chap. 2.

40. Mantena, *Alibis of Empire*. Here, as elsewhere, so-called customary law was not a written code that was then simply "applied" to the cases at hand. Rather, "customary law" needed to be ascertained and produced. For accounts of the process of producing "customary law" and "tradition" at different sites, see Cohn, "From Indian Status to British Contract"; Chanock, *The Making of South African Legal Culture*; Chanock, *Law, Custom, and Social Order*; Mani, *Contentious Traditions*; and Kelly, *A Politics of Virtue*.

41. Mamdani, *Define and Rule*, especially chap. 2.

42. Chanock, *The Making of South African Legal Culture*, 19. "Indians" in "Africa" posed a specific and difficult legal conundrum. The issues are complex and strain accepted ways of thinking about legal jurisdiction, particularly since the different laws did not follow any logical consistency. For a fine analysis of issues of jurisdiction and the portability of personal law, in general, and, more specifically, with regard to how such issues framed debates over Muslim personal law in Fiji, see Koya, "The Campaign for Islamic Law in Fiji."

43. Chanock, *The Making of South African Legal Culture*, 19.

44. This was not only to manage migration literally from outside southern Africa but, more importantly, to manage migration within southern Africa. In the early twentieth century, as the global control over the movement of people was increasingly taking shape along the axis of racialized nationality, a similar process was being replicated within the boundaries of what constituted the territorial South African state. Thus, even as international migration was becoming a federal rather than local matter, monitored through a complex system of passports and visas, in South Africa processes such as rural-urban migration and inter-provincial migration was being strictly controlled through a complex system of passes and licenses. On these different issues, see McKeown, *Melancholy Order*; Chanock, *The Making of South African Legal Culture*; Mamdani, *Citizen and Subject*; and Breckenridge, *Biometric State*.

45. "Fatima *v.* Rex," *Transvaal Leader*, February 15, 1912, Department of Commerce and Industry (Emigration Proceedings—Part A), September 1912, Proceedings No. 8–9, Enclosure 1, Annex 3, NAI.

46. "Fatima *v.* Rex."

47. "Fatima *v.* Rex." See also "Mrs. Jussat's Case," *Indian Opinion*, March 9, 1912, *The Collected Works of Mahatma Gandhi* (hereafter cited as CWMG) 9:243–44.

48. "Fatima *v.* Rex."

49. "Fatima *v.* Rex."

50. London All-India Moslem League to Colonial Office, October 31, 1911, Department of Commerce and Industry (Emigration Proceedings—Part A), September 1912, Proceedings No. 8–9, Enclosure 1, Annex 1, NAI.

51. M. K. Gandhi, "Indian Wives," *Indian Opinion*, July 8, 1911, CWMG, 9:120. For an analysis of Gandhi's publishing activities in South Africa, with significant discussion of the *Indian Opinion*, see Hofmeyr, *Gandhi's Printing Press.*

52. South Africa British Indian Committee to Colonial Office, May 6, 1912, Department of Commerce and Industry (Emigration Proceedings—Part A), September 1912, Proceedings No. 8–9. Enclosure 1, Annex 5, NAI.

53. South Africa British Indian Committee to Colonial Office, May 6, 1912.

54. South Africa British Indian Committee to Colonial Office, May 6, 1912.

55. South Africa British Indian Committee to Colonial Office, May 6, 1912.

56. For details, see Beall, "Women under Indentured Labour"; Sheik, "Colonial Rites"; and Havaldar, "'Civilizing' Marriage."

57. A policy of so-called noninterference also informed the approach of the Natal government with respect to indentured Indians until 1891, when this policy was, effectively, legislatively abandoned.

58. Polygamy, as noted earlier, was "preserved" as "custom" within law governing the "native" population.

59. South Africa British Indian Committee to Colonial Office, June 5, 1912, Department of Commerce and Industry (Emigration Proceedings—Part A), September 1912, Proceedings No. 8–9, Enclosure 2, Annex 1, NAI.

60. India Office, London to Colonial Office, August 16, 1912, Department of Commerce and Industry (Emigration Proceedings—Part A), September 1912, Proceedings No. 8–9, Enclosure 3, NAI.

61. South Africa British Indian Committee to Colonial Office, June 5, 1912, Department of Commerce and Industry (Emigration Proceedings—Part A), September 1912, Proceedings No. 8–9, Enclosure 2, Annex 1, NAI.

62. "Text of the Searle Judgment," appendix I, CWMG, 12:568.

63. "Text of the Searle Judgment," 568–69.

64. The formulation of "concubines" was one Gandhi would use repeatedly. See, for instance, "New Bill," *Indian Opinion*, April 19, 1913, CWMG, 12:36; Gandhi to Secretary for the Interior, September 22, 1913, CWMG, 12:201; and "The Marriage Question," *Indian Opinion*, October 1, 1913, CWMG, 12:225. The feminist

suggestion that marriage constitutes legalized prostitution is here inadvertently, and amusingly, vindicated.

65. An inquiry, under the supervision of Sir William Solomon, hence referred to as the Solomon Commission, was appointed to investigate the final phase of satyagraha and the strike by indentured workers in South Africa and recommend resolutions to the "Indian Question." Gandhi and others objected to the composition of the committee and Gandhi exhorted Indians to boycott the Commission and refuse to appear to give evidence. The call for a boycott was largely, but not entirely, successful. Report of the Union of South Africa Indian Enquiry Commission, Department of Commerce and Industry (Emigration Proceedings—Part A), May 1914, Proceedings No. 3–4, NAI (emphasis added). It would take me too far afield to adequately address the implications of the legal status of children, but let me point out that not only would children of polygamous marriages be denied entry into South Africa, but those already there could be—and were—denied their inheritance. Since, in this instance, the state was largely concerned with the marriage arrangements of propertied men, coding their children as illegitimate effectively meant expropriating the property held by Indians.

66. Examination of Sir Benjamin Robertson, January 29, 1914, Nos. 712A, 723A, 726A, Indian Enquiry Commission, Department of Commerce and Industry (Emigration Proceedings—Part A), April 1914, File No. 24, NAI. It is difficult to ascertain if Robertson refers to forty women or forty men who were in polygamous marriages. But, by his account, there were about forty "such cases."

67. Another significant objection to the bill concerned the restrictions it contained on inter-provincial migration.

68. The career of satyagraha from 1906/1907, when it commenced, to 1913, when the Searle decision was delivered, was one of relatively steady decline in interest and participation. It had initially appealed, by the thousands, to the commercial elite and ex-indentured Indians. But various compromises Gandhi was seen to have made with the government, as well as mass arrests, imprisonment, and financial ruin, had eroded support so that, by 1909, the movement in South Africa was largely somnolent and kept afloat mostly by donations from India and the participation of perhaps one hundred supporters. Until 1913, its demands were also largely restricted to concerns directly affecting the elite and petty bourgeoisie, rather than the indentured population. For an important account of these events that does not construct a presentist hagiography centered on Gandhi, see Swan, *Gandhi*. See also Mongia, "Gender and the Historiography of Gandhian *Satyagraha* in South Africa."

69. "Hindus and Mahomedans Beware," *Indian Opinion*, March 22, 1913, CWMG, 11:496.

70. "Attack on Indian Religions," *Indian Opinion*, March 22, 1913, CWMG, 11:497.

71. A. M. Cachalia to Governor General, "Resolutions at Vrededorp Mass

Meeting," March 30, 1913, Governor General, Despatches and Correspondence, vol. 894, no. 15/375, SAB.

72. "Resolutions at Vrededorp Mass Meeting."

73. "Resolutions at Vrededorp Mass Meeting" (emphasis added).

74. Gandhi to E. F. C. Lane, Private Secretary to General Smuts, Minister of the Interior, April 9, 1913, CWMG, 12:9.

75. "Conversation with Kasturba Gandhi," *Indian Opinion*, October 1, 1913, CWMG, 12:31.

76. "Conversation with Kasturba Gandhi."

77. "Indian Women as Passive Resisters," *Indian Opinion*, May 10, 1913, CWMG, 12:65. The telegram was sent on May 4, 1913.

78. "The Women's Resolution," *Indian Opinion*, May 10, 1913, CWMG, 12:66.

79. "Indian Marriages," *Indian Opinion*, March 29, 1913, CWMG, 11:505.

80. Let us recall that the honor of women had pervaded the larger discourse of Indian opposition to the activities of the state in 1906, when satyagraha commenced and the Transvaal "Black Act" requirement for Indian women to carry identification passes was defeated. The resolution, as noted earlier, was that only men would be required to have identification papers, listing the names of their wife, or wives, and daughters.

81. "Mrs. Pankhurst's Sacrifice," *Indian Opinion*, April 19, 1913, CWMG, 12:37.

82. "Draft Immigration Regulation Bill and the Act [as Gazetted]," appendix VI, CWMG, 12:580. The appendix contains, in two columns, both the bill as introduced and the act as gazetted.

83. A. M. Cachalia to the Secretary of the Interior, September 12, 1913, *Union of South Africa, Correspondence Relating to the Immigrants Regulation Act and Other Matters Affecting Asiatics in South Africa*, Parliamentary Papers (House of Commons) 59, Cd. 7111, 26.

84. The support increased somewhat after September 23, when the first passive resisters, twelve men and four women, courted arrest and were imprisoned for three months. See, for instance, East London [South Africa] British Indian Association to Governor General, Pretoria, September 29, 1913, *Correspondence Relating to the Immigrants Regulation Act and Other Matters Affecting Asiatics in South Africa*, Parliamentary Papers (House of Commons) 59, Cd. 7111, 35; and Awakened India Society, Cape Town, to Governor General, Pretoria, September 26, 1913, Governor General, Despatches and Correspondence, vol. 896, no. 15/478, SAB.

85. It is important that the case came forth in Natal (not in the Transvaal or one of the other provinces) where polygamy had been legislatively illegalized for the Indian indentured population since 1891, embodying a reversal of "noninterference" in the customs of Indians. What the judgment thus indicated was that, in the newly formed Union of South Africa, the most restrictive—rather than the most permissive—policy on Indian marriage between the different provinces would hold validity.

86. South Africa British Indian Committee to Colonial Office, October 6, 1913, *Union of South Africa, Correspondence Relating to the Immigrants Regulation Act and Other Matters Affecting Asiatics in South Africa*, Parliamentary Papers (House of Commons) 59, Cd. 7111, 24–25.

87. Anjuman Islam to Governor General, October 2, 1913, Governor General, Despatches and Correspondence, vol. 896, no. 15/481, SAB.

88. R. G. Teikamdas, Chairman, Port Elizabeth British Indian Association, to Governor General, October 2, 1913, *Union of South Africa. Correspondence Relating to the Immigrants Regulation Act and Other Matters Affecting Asiatics in South Africa*, Parliamentary Papers (House of Commons) 59, Cd. 7111, 44.

89. Indian Political Association, Kimberly to Governor General, October 9, 1913, *Union of South Africa, Correspondence Relating to the Immigrants Regulation Act and Other Matters Affecting Asiatics in South Africa*, Parliamentary Papers (House of Commons) 59, Cd. 7111, 47; V. R. Pillay, on behalf of Indians in Pietermaritzburg, to Governor General, October 8, 1913, Governor General, Despatches and Correspondence, vol. 897, no. 15/488, SAB.

90. "The Marriage Question," *Indian Opinion*, October 1, 1913, CWMG, 12:230.

91. Robert Huttenback, for instance, notes that the Searle judgment, "more than any other factor, was to usher in a renewal of *satyagraha*," but his analysis sees this as a *cause* of satyagraha and does not explore the *form* in which the Searle decision was used to mobilize support. Huttenback, *Gandhi in South Africa*, 307. Maureen Swan mentions the Searle decision and notes that support for passive resistance increased in October; however, in an otherwise meticulously researched and exceedingly well-documented study, she does not trace this increase to the Kulsan Bibi judgment. Swan, *Gandhi*, 236–44.

92. See, especially, Gandhi on the first passive resistance pledge in 1906 where the key term was not nation but God. Gandhi, *Satyagraha in South Africa*, 103–7.

93. In stark contrast, elite trader organizations, such as the Natal Indian Congress, that by 1913 opposed satyagraha, did not even mention the marriage question in their objections to the Immigration Act, focusing instead on issues of domicile and trading licenses. Natal Indian Congress to Governor General, June 30, 1913, Governor General, Despatches and Correspondence, vol. 896, no. 15/449, SAB, and Natal Indian Congress to Governor General, September 8, 1913, Governor General, Despatches and Correspondence, vol. 896, no. 15/468, SAB.

94. The London All-India Moslem League found that the Kulsan Bibi judgment "filled them with amazement" and saw the treatment of Indians in South Africa as "little short of persecution." In their opinion, moreover, the condition was "prejudicial to the best interests of Empire" and the "mere fact" that South Africa was a self-governing colony would not "prevent all the odium of the ill-treatment and injustice [toward Indians] from resting on the Imperial government." London All-India Moslem League to India Office, October 15, 1913, Governor General, Despatches and Correspondence, vol. 897, no. 15/520, SAB, and

Hamidia Islamic Society to Minister of the Interior, March 23, 1914, Governor General, Despatches and Correspondence, vol. 899, no. 15/673, SAB.

95. Report of the Union of South Africa Indian Enquiry Commission, Department of Commerce and Industry (Emigration Proceedings—Part A), May 1914, Proceedings No. 3–4, NAI.

96. Report of the Union of South Africa Indian Enquiry Commission.

97. See Charu Gupta, "'Innocent' Victims/'Guilty' Migrants," for a detailed analysis of how concerns over the sexual morality of migrants entangled with caste pervaded the print culture of the Hindi public sphere at the time. See also Charu Gupta, *The Gender of Caste*.

98. Governor General Gladstone's Minute on Meeting with Gokhale (confidential), November 16, 1912, appendix XXII, *CWMG*, 11:575.

99. Quoted in Tinker, *New System*, 378.

100. For a discussion of some of these issues, see Mongia, "Gender and the Historiography of Gandhian *Satyagraha* in South Africa."

101. A. M. Cachalia to the Secretary of the Interior, September 12, 1913, *Union of South Africa, Correspondence Relating to the Immigrants Regulation Act and Other Matters Affecting Asiatics in South Africa*, Parliamentary Papers (House of Commons) 59, Cd. 7111, 26.

102. "An Official Statement," October 15, 1913, *CWMG*, 12:239–41.

103. Cable to G. K. Gokhale, October 22, 1913, *CWMG*, 12:245.

104. Swan, *Gandhi*, 247.

105. See Swan, *Gandhi*, 251–52, and Gandhi, *Satyagraha in South Africa*, 316.

106. For a more in-depth discussion of some of these issues, see Mongia, "Gender and the Historiography of Gandhian *Satyagraha* in South Africa."

107. In 1917, following the end of the war, the entire issue of the legal recognition of Indian marriages would erupt again. See Governor General, South Africa to Colonial Office, May 25, 1917, Position of Indians in South Africa, Department of Commerce and Industry (Emigration Proceedings—Part A), October 1917, Proceedings No. 15–19 (confidential), NAI.

108. Lord Harcourt, Secretary of State for the Colonies, December 6, 1913, Position of Indians in South Africa, Department of Commerce and Industry (Emigration Proceedings—Part A), April 1914, Serial No. 125, Proceedings No. 9–12 (confidential), NAI.

109. "Note" from Sir Syed Ali Imam to Lord Hardinge, Viceroy of India, February 3, 1914, Validation of Indian Marriages in South Africa, Department of Commerce and Industry (Emigration Proceedings—Part A), April 1914, Proceedings No. 4–8 (confidential), NAI.

110. Comments of the Commission in Examination of Sir Benjamin Robertson, January 29, 1914, Nos. 720A, 722A, Indian Enquiry Commission, Department of Commerce and Industry (Emigration Proceedings—Part A), April 1914, File No. 24, NAI (emphasis added).

111. Chanock, *The Making of South African Legal Culture*, discusses the

importance of marriage regulations to the formation of legal regimes in South Africa.

112. For a related analysis of marriage arrangements in Fiji, see Kelly, "Fear of Culture."

113. The complex and contentious debates over the precise form of such documentation and verification are properly the subject of a separate account. While, beginning in the 1860s, the colonial state in India had attempted to institute a system for the voluntary registration of births, deaths, and marriages, this met with limited success. See Singha, "Colonial Law and Infrastructural Power."

114. On these regulations, see Sheik, "Colonial Rites"; and Havaldar, "'Civilizing' Marriage."

115. See, for instance, the important early essays in Yuval-Davis and Anthias, eds., *Woman/Nation/State*; Yuval-Davis, *Gender and Nation*; and Kaplan, Alarcon, and Moellem, eds., *Between Woman and Nation*.

116. See Noriel, *The French Melting Pot*; Balibar, "The Nation Form"; and Balibar, *We, the People of Europe?*.

117. Balibar, *We, the People of Europe?*, 123.

118. For an analysis of recent debates and legal responses to "forced marriages" of Muslim immigrants and their place within the production of "white Europe," focused particularly on Norway, see Razack, *Casting Out*, chap. 4. There are several resonances between issues addressed in this chapter and recent debates and contestations, at numerous sites, regarding "same-sex marriage" and concerns about their validity across state jurisdictions that, regrettably, I cannot pursue here.

119. Harris, "Gandhi, the Chinese and Passive Resistance."

120. Many studies of nationalism in South Africa have focused, not without good reason, on the development of a white Afrikaner nationalism following the formation of the Union and its confrontation, over the course of the twentieth century, with a pan–South African black nationalism, both of which were directed toward "capturing" the state. We can understand the activities of the Indian population as a "subordinate" nationalism that, while unable to "fill" or "capture" the state, nonetheless did not leave it "empty." For discussions of important aspects of South African nationalism, see Marks and Trapido, eds., *The Politics of Race, Class and Nationalism in Twentieth-Century South Africa*; Hofmeyr, "Building a Nation from Words"; McClintock, *Imperial Leather*. For the entanglements between Indians and Africans and the trajectories of midtwentieth-century nationalism in Natal, see Soske, "'Wash Me Black Again.'"

Chapter 4: Race, Nationality, Mobility

Epigraph: Étienne Balibar, "The Nation Form: History and Ideology," translated by Chris Turner, in *Race, Nation, Class: Ambiguous Identities*, by Étienne Balibar and Immanuel Wallerstein, 86–106 (New York: Verso, 1991).

1. On the lack of systematization of the passport system, see Torpey, *The Invention of the Passport*; for an overview of different kinds or modalities of passports, see Salter, *Rights of Passage.*

2. Gilroy, *"There Ain't No Black in the Union Jack,"* 46.

3. Edward Lawford, Solicitor to the East India Company, to David Hill, June 12, 1838, *Papers Respecting the East India Labourers' Bill*, 2–3, IOR.

4. Edward Lawford, Solicitor to the East India Company, to David Hill, June 12, 1838.

5. Translation of a Report by P. D'Epinay, Procureur-Général, Mauritius, January 5, 1836, Enclosure in no. XIX, *Papers Respecting the East India Labourers' Bill*, 56, IOR.

6. Report by P. D'Epinay.

7. Question whether the term *emigrant* applies to soldiers recruited in India under agreement with the Colonial Secretary for service in Africa, Home Department (Sanitary/Plague), February 1899, Proceedings No. 114–17, NAI. This definition, in fact, had been adopted in Act XIII of 1864, under the guidance of Henry Maine. See *Report by Mr. Geoghegan on Coolie Emigration from India*, Parliamentary Papers (House of Commons) 47, no. 314 (1874): 39.

8. Subject relative to the proposed engagement by the Hong Kong Colony of 150 natives of Madras for the Police of the Colony, Home Department (Public), January 1862, Proceedings No. 1–4, NAI. The definition of "emigrant," "emigrate," etc., had actually remained basically unaltered since Act XIV of 1839.

9. Due to the Indian Plague, "emigration" had been suspended to many destinations; it is thus all the more interesting that the concern with the migration of the soldiers was related not to preventing their movement due to the threat of carrying disease, but to the technicality of whether they should be construed as "emigrants."

10. Question whether the term *emigrant* applies to soldiers recruited in India under agreement with the Colonial Secretary for service in Africa, Home Department (Sanitary/Plague), February 1899, Proceedings No. 114–17, NAI.

11. The extensive annual Emigration Proceedings, published by the Emigration Branch of the Government of British India from 1871, contain no index entries for the term *passport* for thirty-five years until 1905, when three entries appear, all relating to Australia. Between 1905 and 1917, the frequency of these entries multiplies dramatically and always with regard to migration to white-settler colonies or European nations.

12. Leaving aside the checkered and complicated history of the formation and expansion of the Canadian Confederation, in early twentieth-century Canada the sense of imperial identity sutured to Britain was strong as were the particular political arrangements that structured this relation, from the definition of a Canadian subject to ultimate imperial authority over "foreign" affairs. The Canadian citizen as a distinct legal entity emerged in 1947, with then–Prime

Minister Mackenzie King (who features in the events related here) receiving the first citizenship certificate.

13. Copy of telegram (dated November 13, 1905) forwarded from Secretary of State, London, to Viceroy of India, November 19, 1906, Department of Commerce and Industry (Emigration Proceedings—Part A), May 1907, Proceedings No. 7, Serial No. 1, NAI.

14. Memorandum *re* Immigration of Hindoos to Canada, Ottawa, November 2, 1906, Department of Commerce and Industry (Emigration Proceedings—Part A), May 1907, Proceedings No. 7, NAI.

15. Memorandum *re* Immigration of Hindoos to Canada.

16. Memorandum *re* Immigration of Hindoos to Canada. I cannot dwell here on some of the complexities of asserting that the caste system jeopardized employment. For an analysis of how caste emerged as a key colonial category for ordering India, see Ludden, "Orientalist Empiricism"; and Dirks, *Castes of Mind*.

17. Memorandum *re* Immigration of Hindoos to Canada.

18. Balibar, "Is There a 'Neo-Racism'?," 21.

19. N. D. Daru to Under Secretary of State for India, November 19, 1906, Department of Commerce and Industry (Emigration Proceedings—Part A), May 1907, Proceedings No. 7, Serial No. 10, NAI.

20. Colonel Falk Warren to Under Secretary of State for India, November 22, 1906, Department of Commerce and Industry (Emigration Proceedings—Part A), May 1907, Proceedings No. 7, Serial No. 10, NAI.

21. Colonel Falk Warren to Under Secretary of State for India, January 2, 1907, Department of Commerce and Industry (Emigration Proceedings—Part A), May 1907, Proceedings No. 7, Serial No. 14, NAI.

22. Colonel Swayne had been involved in the matter since the initial scheme was to try and redirect the Indians coming to Canada on to Honduras. The scheme came to naught since "the Indians would not go." India Office Memorandum on Indian Immigration into Canada, August 26, 1915, Department of Commerce and Industry (Emigration Proceedings—Part A), October 1915, Proceedings No. 68, Annexure II (confidential, original consultation), NAI.

23. Governor General of Canada to Earl Crewe, Colonial Office, January 7, 1909, Department of Commerce and Industry (Emigration Proceedings—Part A), May 1909, Proceedings No. 11, Serial No. 5, NAI.

24. Under Secretary of State for India to Under Secretary of State, Colonial Office, October 19, 1906, Department of Commerce and Industry (Emigration Proceedings—Part A), May 1907, Proceedings No. 7, Serial No. 5, NAI.

25. Paraphrase of Telegram from Governor General of Canada to Secretary of State for the Colonies, received Colonial Office, November 11, 1907, Department of Commerce and Industry (Emigration Proceedings—Part A), February 1908, Proceedings No. 18–23, NAI.

26. Telegram from Governor General of Canada to Secretary of State for the Colonies.

27. For details on Chinese and Japanese migration, see Chang, "Enforcing Transnational White Solidarity"; and McKeown, *Melancholy Order*.

28. For an examination of the passport as a document of identity in the Indian context, see Singha, "The Great War and a 'Proper' Passport for the Colony." For other important treatments of the development of passports, see Torpey, *The Invention of the Passport*; and Robertson, *The Passport in America*.

29. Telegram from Viceroy of India, Calcutta, to Secretary of State for India, London, January 22, 1908, Department of Commerce and Industry (Emigration Proceedings—Part A), February 1908, Proceedings No. 18–23, Serial No. 16 (confidential), NAI.

30. Due to the pressure mounted on the government, the Morley-Minto reforms, which increased the number of Indian representatives in the legislative councils, were passed in 1909. However, the Indian nationalist demand, at this stage, did not seek to completely sever its ties with the Empire.

31. For a contemporaneous claim of how the situation of emigrants in South Africa had brought the emigration question into prominence, see the previous chapter and Representation of the United Province Congress Committee Regarding the Position of Indians in Canada, Department of Commerce and Industry (Emigration Proceedings—Part A), June 1915, Proceedings No. 1–2, NAI.

32. Telegram from Viceroy of India, Calcutta, to Secretary of State for India, London, January 22, 1908, Department of Commerce and Industry (Emigration Proceedings—Part A), February 1908, Proceedings No. 28, Serial No. 16 (confidential), NAI.

33. Colonel Falk Warren to Under Secretary of State for India, November 22, 1906, Department of Commerce and Industry (Emigration Proceedings—Part A), May 1907, Proceedings No. 7, Serial No. 10, NAI; N. D. Daru to Under Secretary of State for India, November 19, 1906, Department of Commerce and Industry (Emigration Proceedings—Part A), May 1907, Proceedings No. 7, Serial No. 10, NAI.

34. Lord Grey, Governor General of Canada, to Secretary of State for India, September 24, 1907, Department of Commerce and Industry (Emigration Proceedings—Part A), February 1908, Proceedings No. 18–23, NAI. For further elaborations on the form and extent of anti-Asiatic racism, see Ward, *White Canada Forever*; Chang, "Enforcing Transnational White Solidarity"; Kazimi, *Undesirables*; and Johnston, *Voyage of the* Komagata Maru.

35. Report of the Committee of the Privy Council, approved by His Excellency the Governor General on March 2, 1908, Department of Commerce and Industry (Emigration Proceedings—Part A), May 1908, Proceedings No. 6, Serial No. 22, Enclosure No. 9, NAI.

36. Cohn, *Colonialism and Its Forms of Knowledge*, 5.

37. Telegram from Governor General of Canada to Secretary of State for the Colonies, London, January 15, 1908, Department of Commerce and Industry

(Emigration Proceedings—Part A), May 1908, Proceedings No. 6, Serial No. 22, Enclosure No. 3, Annex 1, NAI.

38. Companies did not sell "through tickets" to Indians, but did so to Europeans (Johnston, *Voyage of the* Komagata Maru, 17).

39. Secret agent T. R. E. McInnes to Wilfrid Laurier, March 15, 1908, Department of Commerce and Industry (Emigration Proceedings—Part A), May 1908, Proceedings No. 6, Serial No. 22, Enclosure No. 10, Annex 2, NAI.

40. Secret agent T. R. E. McInnes to Wilfrid Laurier.

41. Secret agent T. R. E. McInnes to Wilfrid Laurier.

42. Secret agent T. R. E. McInnes to Wilfrid Laurier (emphasis in original).

43. On the confounding status of the legality of legal exceptions, see Agamben, *State of Exception*. For an exploration addressing this problematic in a colonial context, see Hussain, *The Jurisprudence of Emergency*.

44. In this regard, court responses to the legal challenges to the executive orders concerning travel and migration, issued by U.S. President Donald J. Trump in early 2017, are noteworthy. On January 27, 2017, Trump issued a much-publicized executive order (No. 13769) titled "Protecting the Nation from Foreign Terrorist Entry into the United States." Among other provisions, the order fundamentally altered the U.S. Refugee Admissions Program and prohibited entry for citizens of seven Muslim-majority states for 90 days, pending a review. On February 9, the order was revoked and another, No. 13780 (retaining the title), which sought to address some of the poor framing of the earlier order, was promulgated. The orders are couched in terms of protecting national security that authorizes the president to prohibit the entry of aliens seen to pose a threat. Both orders were the subject of numerous legal challenges. In judgments against Executive Order No. 13780, the ruling of federal judges in Hawaii and Maryland explicitly draw on numerous statements by Trump that promised a "Muslim ban." As such, the courts found the claims of "national security" unpersuasive and the order as directed against a religious group. These decisions are currently under appeal. It remains to be seen if the judgments will stand or fall in the higher courts. For legal details, see Special Collection: Civil Rights Challenges to Trump Refugee/Visa Order, University of Michigan Law School, Civil Rights Litigation Clearinghouse, available at: https://www.clearinghouse.net/results.php?searchSpecialCollection=44 (accessed May 15, 2017). Both orders were also the subject of extensive press coverage.

45. Governor General of Canada to Earl Crewe, Colonial Office, January 7, 1909, Department of Commerce and Industry (Emigration Proceedings—Part A), May 1909, Proceedings No. 11, Serial No. 5, NAI; Report of the Committee of the Privy Council, approved by His Excellency the Governor General on March 2, 1908, Department of Commerce and Industry (Emigration Proceedings—Part A), May 1908, Proceedings No. 6, Serial No. 22, Enclosure No. 9, NAI.

46. Report of the Committee of the Privy Council.

47. Tinker, *Separate and Unequal*, 29.

48. For example, Governor General of Canada to Earl Crewe, Colonial Office, January 7, 1909, Department of Commerce and Industry (Emigration Proceedings—Part A), May 1909, Proceedings No. 11, Serial No. 5, NAI.

49. Governor General of Canada to Earl Crewe, Colonial Office, January 7, 1909, Department of Commerce and Industry (Emigration Proceedings—Part A), May 1909, Proceedings No. 11, Serial No. 5, NAI. The proceedings overflow with a general conflation and confusion between "Indians," "Hindoos," and "Sikhs," though they also point out that some of the immigrants were "Mohameddan."

50. Governor General of Canada to Earl Crewe.

51. Governor General of Canada to Earl Crewe.

52. For example, Governor General of Canada to Earl Crewe.

53. Governor General of Canada to Earl Crewe (emphasis added).

54. Governor General of Canada to Earl Crewe.

55. Government of India to Viscount Morley, Secretary of State for India, May 20, 1909, Department of Commerce and Industry (Emigration Proceedings—Part A), May 1909, Proceedings No. 11, Serial No. 6 (emphasis added), NAI.

56. British Indian Subjects in Canada to Colonial Office, London, April 24, 1910, Department of Commerce and Industry (Emigration Proceedings—Part A), October 1910, Proceedings No. 47, Serial No. 8, Enclosure No. 1, Annex 1, NAI.

57. British Indian Subjects in Canada to Colonial Office.

58. Observation by W. H. Clark, Government of India, September 20, 1913, Department of Commerce and Industry (Emigration Proceedings—Part A), October 1913, Proceedings No. 29–30, Serial No. 44 (confidential, original consultation), NAI.

59. Enclosure in letter from Under Secretary of State to Governor General's Secretary, April 15, 1912, Department of Commerce and Industry (Emigration Proceedings—Part B), May 1912, Proceedings No. 58 (confidential, original consultation), NAI.

60. Secretary of State for the Colonies to Governor General, Canada, August 17, 1912, Department of Commerce and Industry (Emigration Proceedings—Part B), September 1912, Proceedings No. 14–15 (confidential, original consultation), NAI.

61. Secretary of State for the Colonies to Governor General, Canada.

62. Secretary of State for the Colonies to Governor General, Canada.

63. Secretary of State for the Colonies to Governor General, Canada.

64. Secretary of State for the Colonies to Governor General, Canada, Attached Note.

65. In an analysis of the "Hindu woman's question," debates that concerned the migration of wives of Indian/Sikh men to Canada, Ena Dua shows the centrality of fears of miscegenation and the potential of Asian men to "debauch" white women as also the place of marriage to the production of "ethnic communities." See Dua, "Racializing Imperial Canada" and "The Hindu Woman's Question."

66. See also the discussion of bureaucratic discretion and administrative law in the previous chapter.

67. See Dua, "Racializing Imperial Canada."

68. Memorial from the President of the United Provinces Congress Committee, Allahabad, to the Secretary to the Government of India, January 25, 1915, Department of Commerce and Industry (Emigration Proceedings—Part A), June 1915, Proceedings No. 1–2, NAI.

69. For biographical accounts on revolutionary Indian emigrants in Canada, see Johnston, *Voyage of the* Komagata Maru, and the confidential report prepared by James C. Ker in 1917, published as *Political Trouble in India*.

70. These included the *Free Hindustan*, the *Swadesh Sewak*, and the *Hindusthanee*. The Canadian revolutionaries were also in contact and concert with Har Dayal, the founder of the *Ghadr* ("Anarchy" or "Revolution," often mistranslated as "Mutiny") party in the United States. For details on the connections between the *Ghadr* revolutionaries and the events in Canada, see Puri, *Ghadar Movement*; Ramnath, *Haj to Utopia*; and Sohi, *Echoes of Mutiny*.

71. Comments of S. H. Slater, September 19, 1913, Department of Commerce and Industry (Emigration Proceedings—Part A), October 1913, Proceedings No. 29–30 (confidential, original consultation), NAI.

72. Comments of W. H. Clark, September 20, 1913, Department of Commerce and Industry (Emigration Proceedings—Part A), October 1913, Proceedings No. 29–30 (confidential, original consultation), NAI.

73. Comments of S. H. Slater regarding telegram from Secretary of State, September 19, 1913, Department of Commerce and Industry (Emigration Proceedings—Part A), October 1913, Proceedings No. 29–30 (confidential, original consultation), NAI.

74. Comments of J. F. Gruning regarding telegram from Secretary of State, September 20, 1913, Department of Commerce and Industry (Emigration Proceedings—Part A), October 1913, Proceedings No. 29–30 (confidential, original consultation), NAI.

75. Johnston, *Voyage of the* Komagata Maru, 20.

76. Confidential letter from the Governor General of Canada to Colonial Office, December 31, 1913, Department of Commerce and Industry (Emigration Proceedings—Part A), June 1914, Proceedings No. 10–11, Enclosure No. 2, NAI. For details on the legalities involved, see Walker, *"Race," Rights, and the Law in the Supreme Court of Canada*, 257–58.

77. Confidential letter from the Governor General of Canada to Colonial Office.

78. Johnston, *Voyage of the* Komagata Maru, 29–38.

79. Report of the Committee of the Privy Council, approved February 23, 1914, Department of Commerce and Industry (Emigration Proceedings—Part A), June 1914, Proceedings No. 10–11, Enclosure No. 11, NAI.

80. Unfortunately, it would take me too far afield to present here the extraor-

dinary web of events surrounding the *Komagata Maru* incident, but the following is a bare-bones sketch. The immigration agent at Vancouver, a Malcolm Reid, in direct contravention of the law, at first delayed hearings of the Board of Inquiry. He then held hearings, but withheld any decision, regarding acceptance or deportation of the passengers, so that there would be no decision to challenge in court. Eventually, at the end of June, more than a month after the *Komagata Maru* had arrived in Vancouver, Edward Bird, the lawyer hired to represent the Indians, was allowed to try a "test case" in the Court of Appeal where the case would be decided by a panel of judges. Bird lost the appeal with the panel of judges upholding the Immigration Act and the Orders in Council. However, the matter did not end there, and, ultimately, with an array of several hundred armed militia men lined up on the pier, the *Komagata Maru* was escorted out of the Vancouver harbor on July 23 under the guard of the *Rainbow* (which, along with the *Niobe*, constituted the entire Canadian navy at the time) and the immigration vessel the *Sea Lion*. On their return to India, the passengers were met by the police as seditionists and nineteen were killed in the fracas that followed. Thirty-one were imprisoned, and even those released were closely watched by the police. Twenty-seven, including Sardar Gurdit Singh, were fugitives. In 1922, at the recommendation of Gandhi and the Indian National Congress, Gurdit Singh turned himself in to the police and spent five years in prison. For further details, see Johnston, *Voyage of the* Komagata Maru; and Dhamoon et al., eds., *Charting Imperial Itineraries*.

81. For further details on the legal complexities embodied in and provoked by the *Komagata Maru*, see Mongia, "The *Komagata Maru* as Event." For a recent exploration of the entanglements of law, jurisdiction, and temporality as they related to the *Komagata Maru*, see Mawani, *Across Oceans of Law*.

82. Official Report of a Debate in the Canadian House of Commons on Asiatic Immigration, Department of Commerce and Industry (Emigration Proceedings—Part A), October 1914, Proceedings No. 1, NAI.

83. Official Report of a Debate in the Canadian House of Commons on Asiatic Immigration.

84. India Office Memorandum on Indian Immigration into Canada, August 26, 1915, Department of Commerce and Industry (Emigration Proceedings—Part A), October 1915, Proceedings No. 68 (confidential, original consultation), NAI.

85. Memorial regarding the grievances of Indians in Canada, Department of Commerce and Industry (Emigration Proceedings—Part A), April 1914, Proceedings No. 13–16, NAI.

86. Memorial regarding the grievances of Indians in Canada; Confidential Weekly Diary for the week ending the 13th of June 1914 of the Superintendent of Police, Lahore, Department of Commerce and Industry (Emigration Proceedings—Part A), July 1914, Proceedings No. 3, NAI.

87. Comments of S. H. Slater to R. E. Enthoven, May 26, 1914, Department of

Commerce and Industry (Emigration Proceedings—Part A), September 1914, Proceedings No. 18–20 (confidential, original consultation), NAI.

88. Comments of R. W. Gillian, June 23, 1914, Department of Commerce and Industry (Emigration Proceedings—Part A), September 1914, Proceedings No. 18–20 (confidential, original consultation), NAI.

89. Comments of R. W. Gillian, June 23, 1914.

90. Comments of R. W. Gillian, June 23, 1914.

91. Comments of R. W. Gillian, June 23, 1914.

92. Comments of R. W. Gillian, June 23, 1914; Comments of R. E. Enthoven, June 13, 1914, Department of Commerce and Industry (Emigration Proceedings—Part A), September 1914, Proceedings No. 18–20 (confidential, original consultation; emphasis added), NAI.

93. Fletcher, "'In Exercise of Their Rights of British Citizenship.'"

94. Comments of Lord Hardinge, Viceroy of India, July 8, 1914, Department of Commerce and Industry (Emigration Proceedings—Part A), September 1914, Proceedings No. 18–20 (confidential, original consultation), NAI; Comments of R. E. Enthoven, June 13, 1914, Department of Commerce and Industry (Emigration Proceedings—Part A), September 1914, Proceedings No. 18–20 (confidential, original consultation), NAI. This is akin to what we now know as a visa.

95. Compulsory Passport Regulations, Department of Commerce and Industry (Emigration Proceedings—Part A), June 1917, Proceedings No. 8–22, NAI. For important feminist, anti-racist elaborations of racial formation in Canada, including issues of indigeneity and more contemporary debates, see Bannerji, *Dark Side of the Nation*; Razack, *Dark Threats and White Knights*; Razack, *Casting Out*; and Thobani, *Exalted Subjects*.

96. For an overview of different grounds of state sovereignty, see Barkin, "The Evolution of the Constitution of Sovereignty and the Emergence of Human Rights Norms." For an in-depth elaboration of the centrality of colonialism to the making of sovereignty doctrine and international law, see Anghie, *Imperialism, Sovereignty and the Making of International Law*.

97. While pronouncements of the decline, if not the impending demise, of the nation-state have seen a retreat since the turn of the century, they have by no means been laid to rest. As one influential example of an analysis of globalization and the emergence of post-national identities, see Appadurai, *Modernity at Large*.

98. This position can, of course, be further refined. All manner of citizens/subjects inhabit nation-states across the globe and, despite attempts—ranging from genocide to draconian immigration policy—no state can achieve the pure, homogeneous, "national" population a racist logic would desire.

Epilogue

1. Blyton's work (particularly her deeply problematic Noddy series) has rightly been subjected to substantial critique, and many now deem it inappropriate reading material for children.

2. The incident was covered in a range of leading news sources, including the *New York Times*, the *Globe and Mail*, *The Economist*, and *The Guardian*. See the following for details on the fiasco: MacAskill, "Iroquois Lacrosse Team Cleared to Travel by America—Then Blocked by Britain"; "The Iroquois and Their Passports"; Mick, "Low Tech Passports Ground Iroquois Lacrosse Team"; Thomas Kaplan, "Iroquois Defeated by Passport Dispute."

3. *The Guardian* reported it thus: "Asked why the department had dropped its opposition, the state department spokesman, PJ Crowley said: 'There was flexibility there to grant this kind of one-time waiver given the unique circumstances of this particular trip.'" See MacAskill, "Iroquois Lacrosse Team Cleared to Travel by America—Then Blocked by Britain."

4. This intransigence voided the necessity for Canada to respond to the scenario. See Mick, "Low Tech Passports Ground Iroquois Lacrosse Team."

5. Presumably, Clinton's offer to issue team members emergency U.S. passports covered those conceived of as also U.S. citizens, while those understood as also Canadian citizens would have required Canadian passports.

6. The issue would reemerge with the Iroquois Women's Lacrosse team denied travel on their passports to Scotland in 2015. See DaSilva, "Passport Impasse Keeps Haudenosounee Home," and Laskaris, "Passports Rejected." The documents have now, apparently, been technically upgraded to meet—and exceed—the standards demanded by the United States, the United Kingdom, and Canada. See Hill, "My Six Nation Haudenosounee Passport Is Not a 'Fantasy Document.'" However, in my view, it is unlikely that the documents will be recognized given the implications they embody for Canadian and U.S. sovereignty.

7. Recent work has foregrounded some of these threads in Gandhi's thought while he was active in South Africa. See Desai and Vahed, *The South African Gandhi*. See also Markovits, *The Un-Gandhian Gandhi*; Ramachandra Guha, *Gandhi Before India*; and Arundhati Roy, "The Doctor and the Saint."

8. Torpey, *The Invention of the Passport*, 4–5. Moreover, Torpey adopts a sequential, serial approach that recounts how different European polities would come to monopolize migration control. In my view, a relational approach would yield a more accurate analysis.

9. Partha Chatterjee, *The Nation and Its Fragments*, 20.

10. Partha Chatterjee, *The Nation and Its Fragments*, 20.

11. More recently, Chatterjee has also, if implicitly, extended his notion of the rule of colonial difference to conceptualize the corollary to colonial power: imperial power or empire. The *"imperial prerogative,"* he writes, *"lies in the claim*

to declare the colonial exception" which involves the "determination that the universal principles that apply to relations between sovereign states cannot apply in this exceptional case" (194, emphasis in original). Curiously, for Chatterjee, the imperial prerogative operates at and on the level of state sovereignty and thus only in the inter*national* domain. Thus, "instances of declaring the exception within contexts that are taken to belong to the sphere of the domestic politics of states are not, in this sense, colonial-imperial" (195). It is not clear what is to be gained by posing this restriction and thereby excluding, by definition, the description of practices of "domestic politics" as colonial-imperial. See Chatterjee, *The Black Hole of Empire*.

12. Steinmetz, "The Colonial State as a Social Field" and *The Devil's Handwriting*, especially 27–40.

13. Steinmetz, "The Colonial State as a Social Field," 591 (emphasis in original).

14. Steinmetz, "The Colonial State as a Social Field," 591 (emphasis in original).

15. For a brief exploration of the limits of Weber's formulation of the state with respect to colonialism, see Bhambra, "Comparative Historical Sociology and the State."

16. Steinmetz, "The Colonial State as a Social Field," 593; Steinmetz, *The Devil's Handwriting*, 28.

17. Antony Anghie's detailed exposition of the formulation of key elements of sovereignty doctrine that took shape as resolutions to colonial situations is directly relevant here. Though coproduced, Anghie argues that doctrine took shape as two mutually exclusive forms of sovereignty for the European and non-European world: "Sovereignty for the non-European world is alienation and subordination rather than empowerment. . . . The development of the idea of sovereignty in relation to the non-European world occurs in terms of dispossession, its ability to alienate its lands and rights" (105). In other words, it is at the moment of thus alienating its rights or lands when the entity accedes to the realm of sovereignty, inclusion at the moment of dispossession. Anghie's argument can be related to Giorgio Agamben's notion of the state of exception, though Anghie arrives at his conclusions through an analysis of several important moments in the constitution of international law and sovereignty doctrine. See Anghie, *Imperialism, Sovereignty and the Making of International Law*; Agamben, *State of Exception*.

18. On the multiple forms of borders—territorial, social, and psychic—see, for instance, Balibar, *We, the People of Europe?*.

19. There is now a growing literature exploring these issues. For examples emanating from and analyzing different sites, see Balibar, *We, the People of Europe?*; Crush, "The Dark Side of Democracy"; Dodson, "Locating Xenophobia"; Kapur, *Makeshift Migrants and the Law*; Razack, *Casting Out*.

BIBLIOGRAPHY

_____

Archival Records

INDIA OFFICE LIBRARY AND RECORDS,
BRITISH LIBRARY, LONDON (IOR)
Home Correspondence, 1845–1856.
Papers Respecting the East India Labourers' Bill. London: J. L. Cox and Sons, 1838.
Revenue, Judicial, and Legislative Committee, Miscellaneous Papers. Emigration (General).
Rules for the Guidance of the Protector, Medical Inspector of Emigrants, and the Colonial Emigration Agents at the Port of Calcutta, Under the Provisions of Act XIII of 1864.
Rules Relating to Colonial Emigration from the Port of Calcutta Under Provisions of Act XXI of 1883.
Rules Relating to Colonial Emigration Under Provisions of Act XXI of 1883, Revised Edition, 1892.
Rules Relating to Emigration from the Port of Calcutta Under the Provisions of Act VII of 1871.

NATIONAL ARCHIVES OF INDIA, NEW DELHI (NAI)
Proceedings of the Agriculture, Revenue, and Commerce Department (Emigration).
Proceedings of the Department of Commerce and Industry (Emigration), Part A and Part B.
Proceedings of the Home Department (Public).
Proceedings of the Home Department (Sanitary/Plague).
Proceedings of the Home, Revenue, and Agriculture Department (Emigration).
Proceedings of the Legislative Department. Letters from the Court of Directors of the East India Company.

NATIONAL ARCHIVES OF MAURITIUS, COROMANDAL (NAM)
HA Series
RA Series, Letters Received from Calcutta, 1827–1835.
RC Series, Petitions Relative to Immigration, 1829–1836 and 1837.

RC Series, Petitions Relative to Slavery.
RD Series, Letters Sent by Emigration Committee, 1839–1844.

NATIONAL ARCHIVES REPOSITORY, NATIONAL ARCHIVES AND RECORDS
SERVICE OF SOUTH AFRICA, PRETORIA (SAB)
Governor General, Despatches and Correspondence.

GOVERNMENT PUBLICATIONS, UNITED KINGDOM
Parliamentary Papers (House of Commons)

Correspondence Relating to Return of Coolies from British Guiana to India 35, no.
     404, 1843.
Correspondence Relative to Indian Labor into Mauritius 30, no. 26, 1842.
*Correspondence Relative to the Introduction of Indian Labourers into Mauritius,
     Report of Coms. of Inquiry* 37, no. 331, 1840.
*Despatches from Sir W. Nicolay on Free Labour in Mauritius, and Introduction of
     Indian Labourers* 37, no. 58, 1840.
*Letter from Government of India, February 1841; Report of Indian Law Coms.,
     January 1841, on Slavery in E. Indies* 28, no. 262, 1841.
*Letter from Secretary to Government of India to Committee on Exportation of
     Hill Coolies; Report of Committee and Evidence* 16, no. 45, 1841.
Papers on Exportation of Hill Coolies, from Government of India 16, no. 427, 1841.
*Regulations and Orders by Government of Bengal for Protection of Coolies to and
     from Mauritius* 35, no. 148, 1843.
Report by Mr. Geoghegan on Coolie Emigration from India 47, no. 314, 1874.
*Reports or Despatches of Governor of British Guiana, Respecting Hill Coolies In-
     troduced into the Colony* 34, no. 77, 1840.

Parliamentary Papers (House of Commons): Command Papers

*Papers Relating to Grievances of Her Majesty's Indian Subjects in the South Afri-
     can Republic* 71, C. 7911, 1895.
*Union of South Africa. Correspondence Relating to the Immigrants Regulation
     Act and Other Matters Affecting Asiatics in South Africa* 59, Cd. 7111, 1914.
*Union of South Africa. Correspondence Respecting a Bill to Regulate Immigration
     into the Union of South Africa, with Special Reference to Asiatics* 52, Cd.
     5579, 1911.
*Union of South Africa. Further Correspondence Relating to a Bill to Regulate Im-
     migration into the Union of South Africa; with Special Reference to Asiatics*
     45, Cd. 6940, 1913.

Other Sources

Abrams, Philip. "Notes on the Difficulty of Studying the State (1977)." *Journal of Historical Sociology* 1, no. 1 (March 1988): 58–89.

Agamben, Giorgio. *State of Exception.* Translated by Kevin Attell. Chicago: University of Chicago Press, 2005.

Ahuja, Ravi. "The Origins of Colonial Labor Policy in Late Eighteenth-Century Madras." *International Review of Social History* 44, no. 2 (1999): 159–95.

Allen, Richard B. *Slaves, Freedmen, and Indentured Laborers in Colonial Mauritius.* Cambridge: Cambridge University Press, 1999.

Amrith, Sunil. *Crossing the Bay of Bengal: The Furies of Nature and the Fortunes of Migrants.* Cambridge, MA: Harvard University Press, 2013.

———. "Indians Overseas? Governing Tamil Migration to Malaya, 1870–1941." *Past and Present* 2, no. 208 (August 2010): 231–61.

———. *Migration and Diaspora in Modern Asia.* Cambridge: Cambridge University Press, 2011.

———. "South Indian Migration, c. 1800–1950." In *Globalising Migration History: The Eurasian Experience (16th–21st Centuries),* ed. Leo Lucassen and Jan Lucassen, 122–48. Leiden: Brill, 2014.

Anderson, Benedict. *Imagined Communities: Reflections on the Origin and Spread of Nationalism.* London and New York: Verso, [1983] 1991.

Anderson, Bridget. *Us and Them?: The Dangerous Politics of Immigration Control.* Oxford: Oxford University Press, 2013.

Anderson, Bridget, Nandita Sharma, and Cynthia Wright, eds. "No Borders as a Practical Political Project." Special Issue, *Refuge* 26, no. 2 (2011).

Anderson, Clare. "Convicts and Coolies: Rethinking Indentured Labour in the Nineteenth Century." *Slavery and Abolition: A Journal of Slave and Post-Slave Studies* 30, no. 1 (2009): 93–109.

———. *Convicts in the Indian Ocean: Transportation from South Asia to Mauritius, 1815–1853.* Basingstoke, UK: Macmillan, 2000.

———. "Oscar Mallitte's Andaman Photographs, 1857–8." *History Workshop Journal* 67, no. 1 (2009): 152–72.

Anderson, Michael. "India, 1858–1930: The Illusion of Free Labor." In *Masters, Servants, and Magistrates in Britain and the Empire, 1562–1955,* ed. Douglas Hay and Paul Craven, 422–54. Chapel Hill: University of North Carolina Press, 2004.

Anghie, Antony. *Imperialism, Sovereignty and the Making of International Law.* Cambridge: Cambridge University Press, 2004.

Appadurai, Arjun. *Modernity at Large: Cultural Dimensions of Globalization.* Minneapolis: University of Minnesota Press, 1996.

Archer, Major Edward. "Free Labour versus Slave Labour: A Letter to the Right Hon. S. Lushington, M.P. and the Opponents of Free Labour, Shewing that in their Opposition to Emigration from India to the British Colonies they are Virtually Encouraging the Slave Trade." London: Smith, Elder and Co., 1840.

Arnold, David. *Colonizing the Body: State Medicine and Epidemic Disease in Nineteenth-Century India*. Berkeley: University of California Press, 1993.

Atiyah, P. S. *An Introduction to the Law of Contract*. Oxford: Clarendon Press, 1995.

———. *The Rise and Fall of Freedom of Contract*. Oxford: Oxford University Press, 1979.

Austin, J. L. *How to Do Things with Words: The William James Lectures Delivered at Harvard University in 1955*. Edited by J. O. Urmson and Mirina Sbisa. Cambridge, MA: Harvard University Press, [1962] 1975.

Bahadur, Gaiutra. *Coolie Woman: The Odyssey of Indenture*. Chicago: University of Chicago Press, 2013.

Baker, J. H. "Review of *The Rise and Fall of Freedom of Contract*, by P. S. Atiyah." *Modern Law Review* 43, no. 4 (July 1980): 467–69.

Balibar, Étienne. "Is There a 'Neo-Racism'?" Translated by Chris Turner. In Balibar and Wallerstein, *Race, Nation, Class*, 17–28.

———. "The Nation Form: History and Ideology." Translated by Chris Turner. In Balibar and Wallerstein, *Race, Nation, Class*, 86–106.

———. *We, the People of Europe? Reflections on Transnational Citizenship*. Translated by James Swenson. Princeton, NJ: Princeton University Press, 2004.

Balibar, Étienne, and Immanuel Wallerstein. *Race, Nation, Class: Ambiguous Identities*. New York: Verso, 1991.

Ballentyne, Tony. "Rereading the Archive and Opening up the Nation-State: Colonial Knowledge in South Asia (and Beyond)." In *After the Imperial Turn: Thinking with and Through the Nation*, ed. Antoinette Burton, 102–21. Durham, NC: Duke University Press, 2003.

Ballentyne, Tony, and Antoinette Burton. *Empires and the Reach of the Global, 1870–1945*. Cambridge, MA: Belknap Press of Harvard University Press, 2014.

Bannerji, Himani. *Dark Side of the Nation: Essays on Multiculturalism, Nationalism and Gender*. Toronto: Canadian Scholars' Press, 2000.

Barkin, J. Samuel. "The Evolution of the Constitution of Sovereignty and the Emergence of Human Rights Norms." *Millennium: Journal of International Studies* 27, no. 2 (1998): 229–52.

Basch, Linda, Nina Glick Schiller, and Cristina Szanton Blanc. *Nations Unbound: Transnational Projects, Postcolonial Predicaments, and Deterritorialized Nation-States*. Amsterdam: Gordon and Breach, 1994.

Bates, Crispin, and Marina Carter. "Sirdars as Intermediaries in Nineteenth-Century Indian Ocean Indentured Labour Migration." *Modern Asian Studies* 51, no. 2 (2017): 462–84.

Baxi, Upendra. "'The State's Emissary': The Place of Law in Subaltern Studies."

In *Subaltern Studies: Writings on South Asian History and Society*, vol. 7, ed. Partha Chatterjee and Gyanendra Pandey, 247–64. New Delhi: Oxford University Press, 1992.

Bayly, C. A. *Empire and Information: Intelligence Gathering and Social Communication in India, 1780–1870*. New Delhi: Cambridge University Press, 1999.

Beall, Jo. "Women under Indenture in Colonial Natal." In *Women and Gender in Southern Africa to 1945*, ed. Cherryl Walker, 146–67. Cape Town: David Philip, 1990.

Behal, Rana P. "Coolie Drivers or Benevolent Paternalists? British Tea Planters in Assam and the Indenture Labour System." *Modern Asian Studies* 44, no. 1 (2010): 29–51.

Benton, Lauren. *Law and Colonial Cultures: Legal Regimes in World History, 1400–1900*. Cambridge: Cambridge University Press, 2002.

———. *A Search for Sovereignty: Law and Geography in European Empires, 1400–1900*. Cambridge: Cambridge University Press, 2009.

Bhabha, Homi. "The Other Question: Stereotype, Discrimination and the Discourse of Colonialism." In *The Location of Culture*, 66–84. New York: Routledge, 1994.

Bhambra, Gurminder. "Comparative Historical Sociology and the State: Problems of Method." *Cultural Sociology* 10, no. 3 (2016): 335–51.

Bhana, Surendra, and Joy B. Brian. *Setting Down Roots: Indian Migrants in South Africa, 1860–1911*. Johannesburg: Witwatersrand University Press, 1990.

Birla, Ritu. *Stages of Capital: Law, Culture, and Market Governance in Late Colonial India*. Durham, NC: Duke University Press, 2009.

Bocker, Anita, Kees Groenendijk, Tetty Havinga, and Paul Minderhoud, eds. *Regulation of Migration: International Experiences*. Amsterdam: Het Spinhuis Publishers, 1998.

Bose, Sugata. *A Hundred Horizons: The Indian Ocean in the Age of Global Empire*. Cambridge, MA: Harvard University Press, 2006.

Brass, Tom, and Henry Bernstein. "Introduction: Proletarianisation and Deproletarianisation on the Colonial Plantation." In *Plantations, Proletarians and Peasants in Colonial Asia*, ed. E. V. Daniel, H. Bernstein, and T. Brass, 1–40.

Brass, Tom, and Marcel van der Linden, eds. *Free and Unfree Labour: The Debate Continues*. Bern: Peter Lang: 1997.

Breckenridge, Keith. *Biometric State: The Global Politics of Identification and Surveillance in South Africa, 1850 to the Present*. Cambridge: Cambridge University Press, 2014.

———. "Gandhi's Progressive Disillusionment: Thumbs, Fingers, and the Rejection of Scientific Modernism in *Hind Swaraj*." *Public Culture* 23, no. 2 (April 2011): 331–48.

Brubaker, Rogers. *Nationalism Reframed: Nationhood and the National Question in the New Europe*. Cambridge: Cambridge University Press, 1996.

Burbank, Jane, and Frederick Cooper. *Empires in World History: Power and the Politics of Difference*. Princeton, NJ: Princeton University Press, 2010.

Burton, Antoinette. *Burdens of History: British Feminists, Indian Women, and Imperial Culture, 1865–1915*. Chapel Hill: University of North Carolina Press, 1994.

Butler, Judith. *Bodies That Matter: On the Discursive Limits of "Sex."* New York: Routledge, 1993.

———. *Gender Trouble: Feminism and the Subversion of Identity*. New York: Routledge, 1990.

Caestecker, Frank. "The Changing Modalities of Regulation in International Migration within Continental Europe, 1870–1940." In *Regulation of Migration*, ed. A. Bocker, K. Groenendijk, T. Havinga, and P. Minderhoud, 73–97.

Calavita, Kitty. "U.S. Immigration and Policy Responses: The Limits of Legislation." In *Controlling Immigration: A Global Perspective*, ed. Wayne A. Cornelius, Philip L. Martin, and James F. Hollifield, 55–82. Stanford, CA: Stanford University Press, 1994.

Caplan, Jane, and John Torpey, eds. *Documenting Individual Identity: The Development of State Practices in the Modern World*. Princeton, NJ: Princeton University Press, 2001.

Carter, Marina. *Servants, Sirdars, and Settlers: Indians in Mauritius, 1834–1874*. Delhi: Oxford University Press, 1995.

———. "Strategies of Labour Mobilisation in Colonial India: The Recruitment of Indentured Workers for Mauritius." In *Plantations, Proletarians and Peasants in Colonial Asia*, ed. E. V. Daniel, H. Bernstein, and T. Brass, 229–45.

———. *Voices from Indenture: Experiences of Indian Migrants in the British Empire*. Leicester: Leicester University Press, 1996.

Chakrabarty, Dipesh. *Provincializing Europe: Postcolonial Thought and Historical Difference*. Princeton, NJ: Princeton University Press, 2000.

Chang, Kornel. "Enforcing Transnational White Solidarity: Asian Migration and the Formation of the U.S.-Canadian Boundary." *American Quarterly* 60, no. 3 (2008): 671–96.

Chanock, Martin. *Law, Custom, and Social Order: The Colonial Experience in Malawi and Zambia*. Cambridge: Cambridge University Press, 1985.

———. *The Making of South African Legal Culture, 1902–1936: Fear, Favour, and Prejudice*. Cambridge: Cambridge University Press, 2001.

———. "South Africa, 1841–1924: Race, Contract, and Coercion." In *Masters, Servants, and Magistrates in Britain and the Empire, 1562–1955*, ed. Douglas Hay and Paul Craven, 338–64. Chapel Hill: University of North Carolina Press, 2004.

Chatterjee, Indrani, and Richard M. Eaton, eds. *Slavery and South Asian History*. Bloomington: Indiana University Press, 2006.

Chatterjee, Partha. *The Black Hole of Empire: History of a Global Practice of Power*. Princeton, NJ: Princeton University Press, 2012.

———. *The Nation and Its Fragments: Colonial and Postcolonial Histories.* Princeton, NJ: Princeton University Press, 1993.

———. *Nationalist Thought and the Colonial World: A Derivative Discourse?* Minneapolis: University of Minnesota Press, [1986] 1993.

Clifford, James. *Routes: Travel and Translation in the Late Twentieth Century.* Cambridge, MA: Harvard University Press, 1997.

———. "Traveling Cultures." In *Cultural Studies*, ed. Lawrence Grossberg, Cary Nelson, and Paula A. Treichler, 96–111. New York: Routledge, 1992.

Cohn, Bernard S., ed. *An Anthropologist among the Historians and Other Essays.* New Delhi: Oxford University Press, 1990.

———. "The Census, Social Structure and Objectification in South Asia." In *An Anthropologist among the Historians and Other Essays*, ed. B. S. Cohn, 224–54.

———. *Colonialism and Its Forms of Knowledge: The British in India.* Princeton, NJ: Princeton University Press, 1996.

———. "From Indian Status to British Contract." In *An Anthropologist among the Historians and Other Essays*, ed. B. S. Cohn, 463–82.

Comoroff, John. "Reflections on the Colonial State, in South Africa and Elsewhere: Factions, Fragments, Facts and Fictions." *Social Identities* 4, no. 3 (1998): 321–61.

Cooper, Frederick. *Colonialism in Question: Theory, Knowledge, History.* Berkeley: University of California Press, 2005.

Cooper, Frederick, and Ann Laura Stoler, eds. *Tensions of Empire: Colonial Cultures in a Bourgeois World.* Berkeley: University of California Press, 1997.

Cooper, Frederick, Thomas C. Holt, and Rebecca J. Scott. *Beyond Slavery: Explorations of Race, Labor, and Citizenship in Postemancipation Societies.* Chapel Hill: University of North Carolina Press, 2000.

Crosby, Alfred W. *Ecological Imperialism: The Biological Expansion of Europe, 900–1900.* Cambridge: Cambridge University Press, 2004.

Crush, Jonathan. "The Dark Side of Democracy: Migration, Xenophobia and Human Rights in South Africa." *International Migration* 38, no. 6 (2001): 103–33.

Daniel, E. Valentine, Henry Bernstein, and Tom Brass, eds. *Plantations, Proletarians and Peasants in Colonial Asia.* London: Frank Cass, 1992.

DaSilva, Matt. "Passport Impasse Keeps Haudenosounee Home." *US Lacrosse Magazine*, July 24, 2015. Accessed November 14, 2016. http://uslaxmagazine .com/international/201415/news/072415_passport_impasse_keeps Δi_haudenosaunee_home.

Derrida, Jacques. *Archive Fever: A Freudian Impression.* Translated by Eric Prenowitz. Chicago: University of Chicago Press, 1995.

———. "Signature Event Context." In *Limited Inc.*, edited by Gerald Graff, translated by Jeffery Mehlman and Samuel Weber. Evanston, IL: Northwestern University Press, 1988.

Desai, Ashwin, and Goolam Vahed. *Inside Indian Indenture: A South African*

Story, 1860–1914. Cape Town: Human Sciences Research Council Press, 2010.

———. *The South African Gandhi: Stretcher-Bearer of Empire.* Palo Alto, CA: Stanford University Press, 2015.

Dhamoon, Rita, Davina Bhandar, Renisa Mawani, and Satwinder Bains, eds. *Charting Imperial Itineraries: Unmooring the* Komagata Maru. Vancouver: University of British Columbia Press, forthcoming.

Dirks, Nicholas. *Castes of Mind: Colonialism and the Making of Modern India.* New Delhi: Permanent Black, 2002.

Dodson, Belinda. "Locating Xenophobia: Debate, Discourse, and Everyday Experience in Cape Town, South Africa." *Africa Today* 56, no. 3 (spring 2010): 2–22.

Dorsett, Shaunnagh, and John McLaren, eds. *Legal Histories of the British Empire: Laws, Engagements, and Legacies.* Abingdon, UK: Routledge, 2014.

Dua, Enakshi. "The Hindu Woman's Question: Canadian Nation Building and the Social Construction of Gender for South Asian-Canadian Women." In *Anti-Racist Feminism: Critical Race and Gender Studies*, ed. Agnus Calliste and George Dei, 55–72. Halifax: Fernwood Publishing, 2000.

———. "Racializing Imperial Canada: Indian Women and the Making of Ethnic Communities." In *Sisters or Strangers: Immigrant, Ethnic, and Racialized Women in Canadian History*, ed. Marlene Epp, Franca Iacovetta, and Francis Swyripa, 71–86. Toronto: University of Toronto Press, 2004.

Dummett, Ann. "The Transnational Migration of People Seen from a Natural Law Tradition." In *Free Movement: Ethical Issues in the Transnational Migration of People and of Money*, ed. Brian Barry and Robert E. Goodin, 169–80. University Park: Pennsylvania State University Press, 1992.

Eltis, David. "Seventeenth Century Migration and the Slave Trade: The English Case in Comparative Perspective." In *Migration, Migration History, History: Old Paradigms and New Perspectives*, ed. Jan Lucassen and Leo Lucassen, 87–109. New York: Peter Lang, 1997.

Emmer, P. C., ed. *Colonialism and Migration: Indentured Labour Before and After Slavery.* Dordrecht: Martinus Nijhoff, 1986.

———. "The Great Escape: The Migration of Female Indentured Servants from British India to Surinam, 1873–1916." In *Abolition and Its Aftermath: The Historical Context, 1790–1916*, ed. David Richardson, 245–66. New York: Frank Cass, 1985.

———. "The Meek Hindu: The Recruitment of Indian Indentured Labourers for Service Overseas, 1870–1916." In *Colonialism and Migration*, ed. Emmer, 187–207.

Engerman, Stanley, ed. *Terms of Labor: Slavery, Serfdom, and Free Labor.* Stanford, CA: Stanford University Press, 1999.

Ewald, François. "Norms, Discipline, and the Law." *Representations*, no. 30 (spring 1990): 138–61.

Fabian, Johannes. *Time and the Other: How Anthropology Makes Its Object*. New York: Columbia University Press, [1983] 2014.

Fletcher, Ian Christopher. "'In Exercise of Their Rights of British Citizenship': The *Komagata Maru* and the Paradox of Imperial Citizenship Before the First World War." Paper presented at conference on "Charting Imperial Itineraries, 1914–2014: Unmooring the *Komagata Maru*." University of Victoria, British Columbia, Canada, May 2014.

Foucault, Michel. *Discipline and Punish: The Birth of the Prison*. Translated by Alan Sheridan. New York: Vintage, [1975] 1979.

———. "Nietzsche, Genealogy, History." In *Language, Counter-Memory, Practice*, edited by Donald F. Bouchard, translated by Donald F. Bouchard and Sherry Simon, 139–64. Ithaca, NY: Cornell University Press, 1977.

Gandhi, Mohandas Karamchand. *The Collected Works of Mahatma Gandhi*. 100 vols. Delhi: Publications Division, Ministry of Information and Broadcasting, Government of India, 1958–1994.

———. *Satyagraha in South Africa*. Translated by Valji Govindji Desai. Ahmedabad: Navjivan Publishing, [1928] 1961.

Gellner, Ernest. *Nations and Nationalism*. Ithaca, NY: Cornell University Press, 1983.

Ghosh, Amitav. *Sea of Poppies*. New Delhi: Penguin, 2008.

Gillion, K. L. *Fiji's Indian Migrants: A History to the End of Indenture in 1920*. Melbourne: Oxford University Press, 1962.

Gilroy, Paul. *The Black Atlantic: Modernity and Double Consciousness*. Cambridge, MA: Harvard University Press, 1993.

———. *"There Ain't No Black in the Union Jack": The Cultural Politics of Race and Nation*. Chicago: University of Chicago Press, [1987] 1991.

Glick Schiller, Nina. "Transmigrants and Nation-States: Something Old and Something New in the U.S. Immigrant Experience." In *The Handbook of International Migration*, ed. C. Hirschman, P. Kasinitz, and J. DeWind, 94–119.

Glick Schiller, Nina, and Peggy Levitt. "Haven't We Heard This Somewhere Before? A Substantive View of Transnational Migration Studies by Way of a Reply to Waldinger and Fitzgerald." Princeton University, Woodrow Wilson School of Public and International Affairs, Center for Migration and Development, Working Papers, 2006.

Gordley, James. "Contract, Property, and the Will: The Civil Law and Common Law Tradition." In *The State and Freedom of Contract*, ed. Harry N. Scheiber, 66–88.

———. *The Philosophical Origins of Modern Contract Doctrine*. Oxford: Clarendon Press, 1991.

Green, Nancy, and F. Weil, eds. *Citizenship and Those Who Leave: The Politics of Emigration and Expatriation*. Urbana: University of Illinois Press, 2007.

Grovogui, Siba N'Zatioula. *Sovereigns, Quasi Sovereigns, and Africans: Race and*

Self-Determination in International Law. Minneapolis: University of Minnesota Press, 1996.

Guha, Ramachandra. *Gandhi Before India.* New Delhi: Penguin, 2013.

Guha, Ranajit. "Dominance Without Hegemony and Its Historiography." In *Subaltern Studies: Writings on South Asian History and Society,* vol. 6, ed. Ranajit Guha, 210–309. New Delhi: Oxford University Press, 1989.

Gupta, Akil. *Red Tape: Bureaucracy, Structural Violence, and Poverty in India.* Durham, NC: Duke University Press, 2012.

Gupta, Charu. *The Gender of Caste: Representing Dalits in Print.* Ranikhet, India: Permanent Black, 2016.

———. "'Innocent' Victims/'Guilty' Migrants: Hindi Public Sphere, Caste and Indentured Women in Colonial North India." *Modern Asian Studies* 49, no. 5 (2015): 1345–77.

Hall, Catherine. *Civilizing Subjects: Metropole and Colony in the English Imagination 1830–1867.* Chicago: University of Chicago Press, 2002.

Hall, Catherine, and Sonya Rose, eds. *At Home with the Empire: Metropolitan Culture and the Imperial World.* Cambridge: Cambridge University Press, 2006.

Hall, Stuart. "Cultural Identity and Diaspora." In *Contemporary Postcolonial Theory: A Reader,* ed. Padmini Mongia, 110–21. London: Arnold, 1996.

———. "Notes on Deconstructing the 'Popular.'" In *People's History and Socialist Theory,* ed. Raphael Samuel, 227–40. London: Routledge and Kegan Paul, 1981.

Harris, Karen. "Gandhi, the Chinese and Passive Resistance." In *Gandhi and South Africa: Principles and Politics,* ed. Judith M. Brown and Martin Prozesky, 69–94. New York: St. Martin's, 1996.

———. "Sugar and Gold: Indentured Indian and Chinese Labour in South Africa." *Journal of Social Science* 25, nos. 1/2/3 (2010): 147–58.

Hartman, Saidiya V. *Scenes of Subjection: Terror, Slavery, and Self-Making in Nineteenth-Century America.* New York: Oxford University Press, 1997.

Havaldar, Anika. "'Civilizing' Marriage: British Colonial Regulation of the Marriages of Indian Indentured Laborers in Natal, 1860–1891." Senior thesis, Columbia University, 2015. Accessed November 3, 2016. http://history.columbia.edu/undergraduate/theses/2015%20Theses/2015Havaldar.pdf.

Hay, Douglas, and Paul Craven. "Introduction." In *Masters, Servants, and Magistrates in Britain and the Empire, 1562–1955,* ed. Douglas Hay and Paul Craven, 1–58. Chapel Hill: University of North Carolina Press, 2004.

———, eds. *Masters, Servants, and Magistrates in Britain and the Empire, 1562–1955.* Chapel Hill: University of North Carolina Press, 2004.

Hill, Sid. "My Six Nation Haudenosounee Passport Is Not a 'Fantasy Document.'" *The Guardian,* October 30, 2015. Accessed November 14, 2016. https://www.theguardian.com/commentisfree/2015/oct/30/my-six-nation-haudenosaunee-passport-not-fantasy-document-indigenous-nations.

Hirschman, Charles, Philip Kasinitz, and Josh DeWind, eds. *The Handbook of International Migration: The American Experience*. New York: Russell Sage, 1999.

Hobsbawm, Eric. *Nations and Nationalism since 1780: Programme, Myth, Reality*. Cambridge: Cambridge University Press, 1990.

Hofmeyr, Isabel. "The Black Atlantic Meets the Indian Ocean: Forging New Paradigms of Transnationalism for the Global South—Literary and Cultural Perspectives." *Social Dynamics: A Journal of African Studies* 33, no. 2 (2007): 3–32.

———. "Building a Nation from Words: Afrikaans Language, Literature and Ethnic Identity, 1902–1924." In *The Politics of Race, Class and Nationalism in Twentieth-Century South Africa*, ed. Shula Marks and Stanley Trapido, 95–123.

———. *Gandhi's Printing Press: Experiments in Slow Reading*. Cambridge, MA: Harvard University Press, 2013.

———. "Universalizing the Indian Ocean." *PMLA* 125, no. 3 (2010): 721–29.

Hollifield, James. "The Emerging Migration State." In "Motion in Place/Place in Motion: 21st Century Migration," ed. Toshio Iyotani and Masako Ishii.

Holt, Thomas C. *The Problem of Freedom: Race, Labor, and Politics in Jamaica and Britain, 1832–1938*. Baltimore: Johns Hopkins University Press, 1992.

Horwitz, Morton. "The Historical Foundations of Modern Contract Law." *Harvard Law Review* 87, no. 5 (March 1974): 917–56.

———. *The Transformation of American Law, 1780–1860*. Cambridge, MA: Harvard University Press, 1977.

Hossain, Purba. "Protests at the Colonial Capital: Calcutta and the Global Debates on Indenture, 1836–42." *South Asian Studies* 33, no. 1 (2017): 37–51.

Hussain, Nasser. *The Jurisprudence of Emergency: Colonialism and the Rule of Law*. Ann Arbor: University of Michigan Press, 2003.

Huttenback, Robert A. *Gandhi in South Africa: British Imperialism and the Indian Question, 1860–1914*. Ithaca, NY: Cornell University Press, 1971.

"The Iroquois and Their Passports: Unfair Play." *Economist*, July 22, 2010. Accessed November 14, 2016. http://www.economist.com/node/16643313.

Iyotani, Toshio, and Masako Ishii, eds. "Motion in Place/Place in Motion: 21st Century Migration." *Japan Centre for Area Studies Symposium Series*, no. 22. Osaka: National Museum of Ethnology, 2005.

Jain, Prakash C. *Racial Discrimination against Overseas Indians: A Class Analysis*. New Delhi: Concept Publishing, 1990.

Johnston, Hugh. *The Voyage of the* Komagata Maru: *The Sikh Challenge to Canada's Colour Bar*, rev. ed. Vancouver: University of British Columbia Press, 2014.

Kale, Madhavi. *Fragments of Empire: Capital, Slavery, and Indian Indentured Labor in the British Caribbean*. Philadelphia: University of Pennsylvania Press, 1998.

Kaplan, Caran, Norma Alarcon, and Minoo Moellem, eds. *Between Woman and*

Nation: Nationalisms, Transnational Feminisms, and the State. Durham, NC: Duke University Press, 1999.

Kaplan, Martha. "Panopticon in Poona: An Essay on Foucault and Colonialism." *Cultural Anthropology* 10, no. 1 (1995): 85–98.

Kaplan, Thomas. "Iroquois Defeated by Passport Dispute." *New York Times*, July 16, 2010. Accessed November 14, 2016. http://www.nytimes.com/2010/07/17/sports/17lacrosse.html.

Kapur, Ratna. "Cross-Border Movements and the Law: Renegotiating the Boundaries of Difference." In *Trafficking and Prostitution Reconsidered: New Perspectives on Migration, Sex Work, and Human Rights*, ed. Kamala Kempadoo, 25–41. Boulder, CO: Paradigm Publishers, 2005.

———. *Makeshift Migrants and the Law: Gender, Belonging, and Postcolonial Anxieties*. New Delhi: Routledge, 2010.

Kayaoglu, Turan. "Westphalian Eurocentrism in International Relations Theory." *International Studies Review* 12, no. 2 (June 2010): 193–217.

Kazimi, Ali. *Undesirables: White Canada and the* Komagata Maru. Vancouver: Douglas and McIntyre, 2011.

Kelly, John D. "'Coolie' as a Labour Commodity: Race, Sex, and European Dignity in Colonial Fiji." In *Plantations, Proletarians and Peasants in Colonial Asia*, ed. E. V. Daniel, H. Bernstein, and T. Brass, 246–68.

———. "Fear of Culture: British Regulation of Indian Marriage in Post-Indenture Fiji." *Ethnohistory* 36, no. 4 (1989): 372–91.

———. "Gaze and Grasp: Plantations, Desires, Indentured Indians, and Colonial Law in Fiji." In *Sites of Desire, Economies of Pleasure: Sexualities in Asia and the Pacific*, ed. Lenore Manderson and Margaret Jolly, 72–98. Chicago: University of Chicago Press, 1997.

———. *A Politics of Virtue: Hinduism, Sexuality, and Countercolonial Discourse in Fiji*. Chicago: University of Chicago Press, 1991.

Kelly, John D., and Martha Kaplan. *Represented Communities: Fiji and World Decolonization*. Chicago: University of Chicago Press, 2001.

Kempadoo, Kamala. "Victims and Agents of Crime: The New Crusade against Trafficking." In *Global Lockdown: Race, Gender, and the Prison Industrial Complex*, ed. Julia Sudbury, 35–55. New York: Routledge, 2005.

Ker, James C. *Political Trouble in India*. Delhi: Oriental Publishers, [1917] 1973.

Kim, Jaeeun. "Establishing Identity: Documents, Performance, and Biometric Information in Immigration Proceedings." *Law and Social Inquiry* 36, no. 3 (2011): 760–86.

Kolsky, Elizabeth. "Codification and the Rule of Colonial Difference: Criminal Procedure in British India." *Law and History Review* 23, no. 3 (fall 2005): 631–83.

———. *Colonial Justice in British India: White Violence and the Rule of Law*. Cambridge: Cambridge University Press, 2010.

Koya, Riyad Sadiq. "The Campaign for Islamic Law in Fiji: Comparison, Codification, Application." *Law and History Review* 32, no. 4 (2014): 853–80.

Kumar, Ashutosh. *Coolies of the Empire: Indentured Indians in the Sugar Colonies, 1830–1920*. Delhi: Cambridge University Press, 2017.

———. "Feeding the *Girmitiya*: Food and Drink on Indentured Ships to the Sugar Colonies." *Gastronomica: The Journal of Critical Food Studies* 16, no. 1 (2016): 41–52.

———. "*Naukari*, Networks, and Knowledge: Views of Indenture in Nineteenth-Century North India." *South Asian Studies* 33, no. 1 (2017): 52–67.

Lake, Marilyn. "From Mississippi to Melbourne via Natal: The Invention of the Literacy Test as a Technology for Racial Exclusion." In *Connected Worlds: History in Transnational Perspective*, ed. Ann Curthoys and Marilyn Lake, 209–30. Canberra: Australian National University Press, 2006.

Lake, Marilyn, and Henry Reynolds. *Drawing the Global Color Line: White Men's Countries and the International Challenge of Racial Equality*. Cambridge: Cambridge University Press, 2008.

Lal, Brij V. Girmitiyas: *The Origins of the Fiji Indians*. Lautoka, Fiji: Fiji Institute of Applied Studies, 2004.

———. "Kunti's Cry: Indentured Women on Fiji's Plantations." *Indian Economic and Social History Review* 22, no. 1 (1985): 55–71.

———. "Labouring Men and Nothing More: Some Problems of Indian Indenture in Fiji." In *Indentured Labour in the British Empire 1834–1920*, ed. Kay Saunders, 126–57.

———. "Understanding the Indian Indenture Experience." *South Asia: Journal of South Asian Studies* 21, suppl. 1 (1998): 215–37.

Laskaris, Sam. "Passports Rejected: Haudenosaunee Women's LAX Withdraws from World Championship." *Indian Country*, July 20, 2015. Accessed November 14, 2016. http://indiancountrytodaymedianetwork.com/2015/07/20/passports-rejected-haudenosaunee-womens-lax-withdraws-world-championships-161139.

Levitt, Peggy, and Nina Glick Schiller. "Transnational Perspectives on Migration: Conceptualizing Simultaneity." *International Migration Review* 38, no. 3 (2004): 1002–40.

Lieberman, David. "Contract Before 'Freedom of Contract.'" In *The State and Freedom of Contract*, ed. Henry N. Scheiber, 89–121.

Liu, Lydia H. "Legislating the Universal: The Circulation of International Law in the Nineteenth Century." In *Tokens of Exchange: The Problems of Translation in Global Circulations*, ed. Lydia H. Liu, 127–64. Durham, NC: Duke University Press, 1999.

Locke, John. *The Second Treatise of Government*. Edited by Thomas Peardon. New York: Macmillan, [1690] 1952.

Look Lai, Walton. *Indentured Labor, Caribbean Sugar: Chinese and Indian Mi-*

grants to the British West Indies, 1838–1918. Baltimore: Johns Hopkins University Press, 1993.

Lowe, Lisa. *The Intimacies of Four Continents*. Durham, NC: Duke University Press, 2015.

Lucassen, Jan, and Leo Lucassen, eds. *Globalising Migration History: The Eurasian Experience (16th–21st Centuries)*. Leiden: Brill, 2014.

Lucassen, Leo. "The Great War and the Origins of Migration Control in Western Europe and the United States (1880–1920)." In *Regulation of Migration*, ed. A. Bocker, K. Groenendijk, T. Havinga, and P. Minderhoud, 45–72.

Ludden, David. "Orientalist Empiricism: Transformations of Colonial Knowledge." In *Orientalism and the Postcolonial Predicament*, ed. Carol A. Breckenridge and Peter Van der Veer, 250–78. Philadelphia: University of Pennsylvania Press, 1993.

Luibhéid, Eithne. *Entry Denied: Controlling Sexuality at the Border*. Minneapolis: University of Minnesota Press, 2002.

MacAskill, Ewen. "Iroquois Lacrosse Team Cleared to Travel by America—Then Blocked by Britain." *The Guardian*, July 15, 2010. Accessed November 14, 2016. https://www.theguardian.com/world/2010/jul/15/iroquois-lacrosse-team-passports-visa-us-uk.

Macpherson, C. B. *The Political Theory of Possessive Individualism*. Oxford: Oxford University Press, [1962] 1979.

Mahase, Radica. "'Plenty a dem run away': Resistance by Indian Indentured Labourers in Trinidad, 1870–1920." *Labor History* 49, no. 4 (2008): 465–80.

Major, Andrea. *Slavery, Abolitionism and the Empire in India, 1772–1843*. Liverpool: Liverpool University Press, 2012.

Mamdani, Mahmood. *Citizen and Subject: Contemporary Africa and the Legacy of Late Colonialism*. Princeton, NJ: Princeton University Press, 1996.

———. *Define and Rule: Native as Political Identity*. Cambridge, MA: Harvard University Press, 2012.

Mani, Lata. *Contentious Traditions: The Debate on Sati in Colonial India*. Berkeley: University of California Press, 1998.

Mantena, Karuna. *Alibis of Empire: Henry Maine and the Ends of Liberal Imperialism*. Princeton, NJ: Princeton University Press, 2010.

Markovits, Claude. *The Global World of Indian Merchants, 1750–1947: Traders of Sind from Bukhara to Panama*. Cambridge: Cambridge University Press, 2000.

———. *The Un-Gandhian Gandhi: The Life and Afterlife of the Mahatma*. New York: Anthem Press, 2004.

Marks, Shula, and Stanley Trapido, eds. *The Politics of Race, Class and Nationalism in Twentieth-Century South Africa*. London: Longman, 1987.

Marx, Karl. *Capital: A Critique of Political Economy*, vol. 1. Edited by Frederick Engels. Translated by Samuel Moore and Edward Aveling. New York: International Publishers, [1867] 1992.

————. *Grundrisse: Foundations of the Critique of Political Economy.* Translated by Martin Nicolaus. New York: Penguin, 1993.

Mawani, Renisa. *Across Oceans of Law: The* Komagata Maru *and Jurisdiction in the Time of Empire.* Durham, NC: Duke University Press, forthcoming.

————. *Colonial Proximities: Crossracial Encounters and Juridical Truths in British Columbia, 1871–1921.* Vancouver: University of British Columbia Press, 2010.

McClintock, Ann. *Imperial Leather: Race, Gender and Sexuality in the Colonial Contest.* New York: Routledge, 1995.

McKeown, Adam. "Global Migration, 1846–1940." *Journal of World History* 15, no. 2 (June 2004): 155–89.

————. *Melancholy Order: Asian Migration and the Globalization of Borders.* New York: Columbia University Press, 2008.

Mehta, Uday Singh. *Liberalism and Empire: A Study in Nineteenth-Century British Liberal Thought.* Chicago: University of Chicago Press, 1999.

Metcalf, Thomas. *Imperial Connections: India in the Indian Ocean Arena, 1860–1920.* Berkeley: University of California Press, 2007.

Mick, Haley. "Low Tech Passports Ground Iroquois Lacrosse Team." *Globe and Mail,* July 14, 2010. Accessed November 14, 2016. http://www.theglobeand mail.com/news/world/low-tech-passports-ground-iroquois-lacrosse-team /article1387213/.

Mintz, Sidney. *Sweetness and Power: The Place of Sugar in Modern History.* New York: Viking-Penguin, 1985.

Mishra, Amit Kumar. "Indian Indentured Labourers in Mauritius: Reassessing the 'New System of Slavery' *vs* Free Labour Debate." *Studies in History* 25, no. 2 (2009): 229–51.

Mohapatra, Prabhu. "Assam and the West Indies, 1860–1920: Immobilizing Plantation Labor." In *Masters, Servants, and Magistrates in Britain and the Empire, 1562–1955,* ed. Douglas Hay and Paul Craven, 455–80. Chapel Hill: University of North Carolina Press, 2004.

————. "The Hosay Massacre of 1884: Class and Community among Indian Immigrant Labourers in Trinidad." In *Work and Social Change in Asia: Essays in Honour of Jan Breman,* ed. Arvind N. Das and Marcel van der Linden, 187–230. New Delhi: Manohar, 2003.

————. "'Restoring the Family': Wife Murders and the Making of a Sexual Contract for Indian Immigrant Labour in the British Caribbean Colonies, 1860–1920." *Studies in History* 11, no. 2 (1995): 227–60.

Mongia, Radhika. "Gender and the Historiography of Gandhian *Satyagraha* in South Africa." *Gender and History* 18, no. 1 (April 2006): 130–49.

————. "Historicizing State Sovereignty: Inequality and the Form of Equivalence." *Comparative Studies in Society and History* 49, no. 2 (2007): 384–411.

————. "Impartial Regimes of Truth: Indentured Indian Labour and the Status of the Inquiry." *Cultural Studies* 18, no. 5 (September 2004): 749–68.

———. "Interrogating Critiques of Methodological Nationalism: Propositions for New Methodologies." In *Beyond Methodological Nationalism: Research Methodologies for Cross-Border Studies*, ed. Anna Amelina, Devrimsel D. Nergiz, Thomas Faist, and Nina Glick Schiller, 198–215. New York: Routledge, 2012.

———. "The *Komagata Maru* as Event: Legal Transformations in Migration Regimes." In *Charting Imperial Itineraries: Unmooring the* Komagata Maru, ed. Rita Dhamoon et al.

———. "Race, Nationality, Mobility: A History of the Passport." *Public Culture* 11, no. 3 (1999): 527–56.

Moses, Jonathon W. *Emigration and Political Development*. New York: Cambridge University Press, 2011.

Mullan, Brendan. "The Regulation of International Migration: The US and Western Europe in Historical Comparative Perspective." In *Regulation of Migration*, ed. A. Bocker, K. Groenendijk, T. Havinga, and P. Minderhoud, 27–44.

Nijhawan, Shobna. "Fallen Through the Nationalist and Feminist Grids of Analysis: Political Campaigning of Indian Women Against Indentured Labour Migration." *Indian Journal of Gender Studies* 21, no. 1 (2014): 111–33.

Niranjana, Tejaswani. *Mobilizing India: Women, Music, and Migration Between India and Trinidad*. Durham, NC: Duke University Press, 2006.

Noriel, Gerard. *The French Melting Pot*. Minneapolis: University of Minnesota Press, 1996.

North-Coombes, M. D. "From Slavery to Indenture: Forced Labour in the Political Economy of Mauritius, 1834–1867." In *Indentured Labour in the British Empire 1834–1920*, ed. Kay Saunders, 78–125.

———. *Studies in the Political Economy of Mauritius*. Moka, Mauritius: Mahatma Gandhi Institute, 2000.

Northrup, David. *Indentured Labor in the Age of Imperialism, 1834–1922*. Cambridge: Cambridge University Press, 1995.

Osiander, Andreas. "Sovereignty, International Relations, and the Westphalian Myth." *International Organization* 55, no. 2 (spring 2001): 251–87.

Pachai, B. *The International Aspects of the South African Indian Question, 1860–1971*. Cape Town: C. Struik, 1971.

Plender, Richard. *International Migration Law*. Leiden: A. W. Sijthoff, 1972.

Portes, Alejandro, and Josh DeWind. "A Cross-Atlantic Dialogue: The Progress of Research and Theory in the Study of International Migration." In *Rethinking Migration: New Theoretical and Empirical Perspectives*, ed. Alejandro Portes and Josh DeWind, 3–28. New York: Berghahn, 2007.

Portes, Alejandro, Luis E. Guarnizo, and Patricia Landolt. "The Study of Transnationalism: Pitfalls and Promise of an Emergent Research Field." *Ethnic and Racial Studies* 22, no. 2 (1999): 217–37.

Prakash, Gyan. *Another Reason: Science and the Imagination of Modern India*. Princeton, NJ: Princeton University Press, 1999.

———. "Body Politic in Colonial India." In *Questions of Modernity*, ed. Timothy Mitchell, 189–223. Minneapolis: University of Minnesota Press, 2000.

———. *Bonded Histories: Genealogies of Labor Servitude in Colonial India*. Cambridge: Cambridge University Press, 1992.

———. "The Colonial Genealogy of Society: Community and Political Modernity in India." In *The Social in Question: New Bearings in History and the Social Sciences*, ed. Patrick Joyce, 81–96. London: Routledge, 2002.

———. "Colonialism, Capitalism, and the Discourse of Freedom." In *"Peripheral" Labour? Studies in the History of Partial Proletarianization*, ed. Shahid Amin and Marcel van der Linden, 9–26. Cambridge: Cambridge University Press, 1997.

Pratt, Geraldine, and Brenda Yeoh. "Transnational (Counter) Topographies." *Gender, Place, and Culture* 10, no. 2 (2003): 159–66.

Puri, Harish K. *Ghadar Movement: Ideology, Organisation and Strategy*, rev. ed. Amritsar: Guru Nanak Dev University, 1993.

Raman, Bhavani. *Document Raj: Writing and Scribes in Early Colonial South India*. Chicago: University of Chicago Press, 2012.

Ramasamy, P. "Labour Control and Labour Resistance in the Plantations of Colonial Malaya." In *Plantations, Proletarians and Peasants in Colonial Asia*, ed. E. V. Daniel, H. Bernstein, and T. Brass, 87–105.

Ramnath, Maia. *Haj to Utopia: How the Ghadar Movement Charted Global Radicalism and Attempted to Overthrow the British Empire*. Berkeley: University of California Press, 2011.

Razack, Sherene. *Casting Out: The Eviction of Muslims from Western Law and Politics*. Toronto: University of Toronto Press, 2008.

———. *Dark Threats and White Knights: The Somalia Affair, Peacekeeping, and the New Imperialism*. Toronto: University of Toronto Press, 2004.

Reddock, Rhoda. "Freedom Denied: Indian Women and Indentureship in Trinidad and Tobago, 1845–1917." In *Review of Women Studies, Economic and Political Weekly* 20, no. 43 (October 26, 1985): WS 79–87.

———. *Women, Labour and Politics in Trinidad and Tobago: A History*. London: Zed Books, 1994.

Richardson, Peter. "Chinese Indentured Labour in the Transvaal Gold Mining Industry, 1904–1910." In *Indentured Labour in the British Empire 1834–1920*, ed. Kay Saunders, 260–90.

———. "Coolies, Peasants, and Proletarians: The Origins of Chinese Indentured Labour in South Africa, 1904–1907." In *International Labour Migration*, ed. Shula Marks and Peter Richardson, 167–85. Middlesex, UK: Maurice Temple Smith, 1984.

Richmond, Theophilus. *The First Crossing, Being the Diary of Theophilus Richmond, Ship's Surgeon Aboard the Hesperus, 1837–38*. Edited by David Dabydeen, Jonathan Morley, Brinsley Samaroo, Amar Wahab, and Brigid Wells. Trinidad: The Caribbean Press, 2010.

Robertson, Craig. *The Passport in America: The History of a Document*. Oxford: Oxford University Press, 2010.

Roy, Arundhati. "The Doctor and the Saint: Ambedkar, Gandhi and the Battle against Caste." *The Caravan: A Journal of Politics and Culture*, March 1, 2014. Accessed November 15, 2016. http://www.caravanmagazine.in/essay/doctor-and-saint.

Roy, Tirthankar. "Sardars, Jobbers, Kanganies: The Labour Contractor and Indian Economic History." *Modern Asian Studies* 42, no. 5 (2008): 971–98.

Salter, Mark. *Rights of Passage: The Passport in International Relations*. Boulder, CO: Lynne Rienner, 2003.

Sanadhya, Totaram. *My Twenty-One Years in the Fiji Islands*. Translated by John D. Kelly and Uttra Kumari Singh. Suva: Fiji Museum, [1914] 2003.

Sandhu, Kernail Singh. *Indian in Malaya: Some Aspects of Their Immigration and Settlement, 1786–1957*. Cambridge: Cambridge University Press, 1969.

Sartori, Andrew. *Liberalism in Empire: An Alternative History*. Oakland: University of California Press, 2014.

Saunders, Kay, ed. *Indentured Labour in the British Empire 1834–1920*. London: Croom Helm, 1984.

Scheiber, Harry N., ed. *The State and Freedom of Contract*. Stanford, CA: Stanford University Press, 1998.

Schuler, Monica. "The Recruitment of African Indentured Labourers for European Colonies in the Nineteenth Century." In *Colonialism and Migration*, ed. P. C. Emmer, 125–61.

Sen, Samita. "Commercial Recruiting and Informal Intermediation: Debate over the Sardari System in Assam Tea Plantations, 1860–1900." *Modern Asian Studies* 44, no. 1 (2010): 3–28.

Sekula, Alan. "The Body and the Archive." *October* 39 (winter 1986): 3–64.

Sewell, William H., Jr. "Three Temporalities: Toward an Eventful Sociology." In *The Historic Turn in the Human Sciences*, ed. Terence McDonald, 245–80. Ann Arbor: University of Michigan Press, 1996.

Sharma, Nandita. *Home Economics: Nationalism and the Making of "Migrant Workers" in Canada*. Toronto: University of Toronto Press, 2006.

———. "On Being *Not* Canadian: The Social Organization of Migrant Workers in Canada." *Canadian Review of Sociology and Anthropology* 38, no. 4 (2001): 415–39.

———. "Travel Agency: A Critique of Anti-Trafficking Campaigns." *Refuge* 21, no. 3 (May 2003): 53–65.

Sheik, Nafisa Essop. "Colonial Rites: Custom, Marriage Law and the Making of Difference in Natal, 1830s–c. 1910." PhD diss., University of Michigan, 2012. Accessed November 3, 2016. http://deepblue.lib.umich.edu/bitstream/handle/2027.42/93815/nsheik_1.pdf?sequence=1.

Shepard, Verene. *Maharani's Misery: Narratives of a Passage from India to the Caribbean*. Mona, Jamaica: University of the West Indies Press, 2002.

Simpson, A. W. B. "The Horwitz Thesis and the History of Contract." *University of Chicago Law Review* 46, no. 3 (spring 1979): 533–601.

Singer, Brian C. J., and Lorna Weir. "Politics and Sovereign Power: Considerations on Foucault." *European Journal of Social Theory* 9, no. 4 (2006): 443–65.

Singha, Radhika. "Colonial Law and Infrastructural Power: Reconstructing Community, Locating the Female Subject." *Studies in History* 19, no. 1 (2003): 88–126.

———. "The Great War and a 'Proper' Passport for the Colony: Border-Crossing in British India, c. 1882–1922." *The Indian Economic and Social History Review* 50, no. 3 (2013): 289–315.

———. "Settle, Mobilize, Verify: Identification Practices in Colonial India." *Studies in History* 16, no. 2 (2000): 151–98.

Sinha, Mrinalini. *Colonial Masculinities: The "Manly Englishman" and the "Effeminate Bengali" in the Late Nineteenth Century.* Manchester, UK: University of Manchester Press, 1995.

———. "Premonitions of the Past." *Journal of Asian Studies* 74, no. 4 (2015): 821–41.

———. *Specters of Mother India: The Global Restructuring of an Empire.* Durham, NC: Duke University Press, 2006.

———. "The Strange Death of an Imperial Ideal: The Case of *Civis Britannicus.*" In *Modernity in South Asia: Modern Makeovers*, ed. Saurabh Dube, 29–42. Oxford: Oxford University Press, 2011.

———. "Totaram Sanadhya's *Fiji Mein Mere Ekkis Varsh*: A History of Empire and Nation in a Minor Key." In *Ten Books That Shaped the British Empire: Creating an Imperial Commons*, ed. Antoinette Burton and Isabel Hofmeyr, 168–89. Durham, NC: Duke University Press, 2014.

———. "Whatever Happened to the Third British Empire? Empire, Nation, Redux." In *Writing Imperial Histories*, ed. Andrew Thompson, 168–87. Manchester, UK: Manchester University Press, 2013.

Smith, Anthony. *The Ethnic Origins of Nations.* Oxford: Blackwell, 1986.

Smith, Richard Saumarez. *Rule by Records: Land Registration and Village Custom in Early British Panjab.* Delhi: Oxford University Press, 1996.

Sohi, Seema. *Echoes of Mutiny: Race, Surveillance, and Indian Anticolonialism in North America.* New York: Oxford University Press, 2014.

Solow, Barbara Lewis, and Stanley L. Engerman, eds. *British Capitalism and Caribbean Slavery: The Legacy of Eric Williams.* Cambridge: Cambridge University Press, 1987.

Soske, Jon. "'Wash Me Black Again': African Nationalism, the Indian Diaspora, and Kwa-Zulu Natal, 1944–1960." PhD diss., University of Toronto, 2009. Accessed November 3, 2016. http://citeseerx.ist.psu.edu/viewdoc/download?doi=10.1.1.471.9143&rep=rep1&type=pdf.

Stanziani, Alessandro. *Bondage: Labor and Rights in Eurasia from the Sixteenth to the Early Twentieth Centuries.* New York: Berghahn, 2014.

———. "Local Bondage in Global Economies: Servants, Wage-Earners, and In-
dentured Migrants in France, Great Britain, and the Mascareigne Islands."
Paper presented at the 8th International Conference on Labour History,
NOIDA, India, March, 2010.

Steinfeld, Robert J. *Coercion, Contract, and Free Labor in the Nineteenth Century.* Cambridge: Cambridge University Press, 2001.

Steinmetz, George. "The Colonial State as a Social Field: Ethnographic Capital
and Native Policy in the German Overseas Empire Before 1914." *American
Sociological Review* 73, no. 4 (August 2008): 589–612.

———. *The Devil's Handwriting: Precoloniality and the German Colonial State
in Qingdao, Samoa, and South West Africa.* Chicago: University of Chicago
Press, 2007.

Stern, Philip J. *The Company-State: Corporate Sovereignty and the Early Modern
Foundations of the British Empire in India.* Oxford: Oxford University Press,
2011.

Stokes, Eric. *The English Utilitarians and India.* Oxford: Clarendon Press, 1959.

Stoler, Ann Laura, ed. *Haunted by Empire: Geographies of Intimacy in North
American History.* Durham, NC: Duke University Press, 2006.

———. *Race and the Education of Desire: Foucault's History of Sexuality and the
Colonial Order of Things.* Durham, NC: Duke University Press, 1995.

Sturman, Rachel. "Indian Indentured Labor and the History of International
Rights Regimes." *American Historical Review* 119, no. 5 (December 2014):
1439–65.

Subrahmanyam, Sanjay. "Connected Histories: Notes Towards a Reconfiguration
of Early Modern Eurasia." *Modern Asian Studies* 31, no. 3 (July 1997): 735–62.

———. *Explorations in Connected History: Mughals and Franks.* Delhi: Oxford
University Press, 2004.

Swan, Maureen. *Gandhi: The South African Experience.* Johannesburg: Raven, 1985.

———. "Ideology in Organized Indian Politics, 1891–1948." In *The Politics of
Race, Class and Nationalism in Twentieth-Century South Africa,* ed. Shula
Marks and Stanley Trapido, 182–208.

Teschke, Benno. *The Myth of 1648: Class, Geopolitics, and the Making of Modern
International Relations.* New York: Verso, 2003.

Thistlethwaite, Frank. "Migration from Europe Overseas in the Nineteenth and
Twentieth Centuries." In *A Century of European Migrations 1830–1930,* ed.
Rudolph Vecoli and Suzanne Senke, 17–49. Urbana: University of Illinois
Press, 1991.

Thobani, Sunera. *Exalted Subjects: Studies in the Making of Race and Nation in
Canada.* Toronto: University of Toronto Press, 2007.

Tinker, Hugh. *A New System of Slavery: The Export of Indian Labour Overseas,
1830–1920.* London: Oxford University Press, 1974.

———. *Separate and Unequal: India and the Indians in the British Commonwealth, 1920–1950.* Vancouver: University of British Columbia Press, 1976.

Torpey, John. *The Invention of the Passport: Surveillance, Citizenship and the State*. Cambridge: Cambridge University Press, 2000.

Trotman, David. *Crime in Trinidad: Conflict and Control in a Plantation Society, 1838–1900*. Knoxville: University of Tennessee Press, 1986.

Trouillot, Michel-Rolph. *Silencing the Past: Power and the Production of History*. Boston: Beacon Press, 1995.

Turner, Mary. "The British Caribbean, 1823–1838: The Transition from Slave to Free Legal Status." In *Masters, Servants, and Magistrates in Britain and the Empire, 1562–1955*, ed. Douglas Hay and Paul Craven, 303–22. Chapel Hill: University of North Carolina Press, 2004.

von Mehren, A. T., and James Gordley. *The Civil Law System: An Introduction to the Comparative Study of Law*. Boston: Little, Brown, 1977.

Wahab, Amar. "In the Name of Reason: Colonial Liberalism and the Government of West Indian Indentureship." *Journal of Historical Sociology* 24, no. 2 (2011): 209–34.

Waldinger, Ralph, and David Fitzgerald. "Transnationalism in Question." *American Journal of Sociology* 109, no. 5 (2004): 1177–95.

Walker, James. *"Race," Rights, and the Law in the Supreme Court of Canada: Historical Case Studies*. Toronto: Osgoode Society for Canadian Legal History and Wilfrid Laurier Press, 1997.

Ward, Peter. *White Canada Forever: Popular Attitudes and Public Policy Toward Orientals in British Columbia*. Montreal: McGill-Queen's University Press, 1978.

Williams, Eric. *Capitalism and Slavery*. Chapel Hill: University of North Carolina Press, 1944.

Wright, Barry. "Macaulay's India Law Reforms and Labor in the British Empire." In *Legal Histories of the British Empire*, ed. Shaunnagh Dorsett and John McLaren, 218–33.

Wong, Diana. "The Rumour of Trafficking and the Management of Migration Studies." In "Motion in Place/Place in Motion: 21st Century Migration," ed. Toshio Iyotani and Masako Ishii, 119–31.

Young, Robert. *White Mythologies: Writing History and the West*. New York: Routledge, 1990.

Yuval-Davis, Nira. *Gender and Nation*. London: Sage, 1997.

Yuval-Davis, Nira, and Floya Anthias, eds. *Woman/Nation/State*. London: Macmillan, 1989.

Zolberg, Aristide. "Matters of State: Theorizing Immigration Policy." In *The Handbook of International Migration*, ed. C. Hirschman, P. Kasinitz, and J. DeWind, 71–93.

———. *A Nation by Design: Immigration Policy in the Fashioning of America*. Cambridge, MA: Harvard University Press, and New York: Russell Sage Foundation, 2006.

INDEX

abolition of slavery: British Slavery Abolition Act (1833) and, 24–25; freedom and, 16, 19, 22–24, 38–39, 48, 168n105; indenture and, 22–26, 30–31, 48, 114, 155n31; migration control and, 32–33, 67–70. *See also* contract; migration (indentured)

Abrams, Philip, 7, 154n24

administrative law, 90, 124, 179n20. *See also* bureaucratic discretion

Agamben, Giorgio, 157n42, 177n92, 197n17

Ali Imam, Syed, 108

American Civil War, 88

Amrith, Sunil, 176n83

Ancient Law (Maine), 60

Anderson, Charles, 40, 45–46, 50–51, 167n97

Anderson, Michael, 166n80

Anghie, Antony, 14, 197n17

Anglo-Boer war. *See* South African war

Anti-Asiatic Leagues, 121–22

anti-colonialism, 67, 85, 104–5, 120–21, 136, 144

apprentices. *See* apprenticeship

apprenticeship, 24–25, 30–31, 33, 160n7

Archer, Edward, 34, 163n48, 169n109

Argentina, 116

Asiatics, 89–92, 110, 121–22, 134

Austin, J. L., 48, 168n105

Australia, 2, 9, 14, 116, 135, 188n11

Balibar, Étienne, 110, 112, 118, 154n22, 197n18

Ballantyne, Tony, 177n2

Behal, Rana, 176n83

Bellingham, Washington, 121

Bentham, Jeremy, 18

Bibi, Kulsan, 102–3, 106, 185n91, 185n94

Bird, Edward, 193n80

Black Act (Transvaal, 1907), 92–93, 95–96, 107, 180n35, 184n80

Blyton, Enid, 141, 196n1

Boer territory. *See* Transvaal

Boer War. *See* South African war

Bombay, 57, 115

Bose, Sugata, 164n63

British Columbia. *See* Canada

British Emancipator, 32

British Empire. *See* British subjects; East India Company; *specific colonies*

British Guiana, 25, 32–33, 54, 58–59, 69, 81, 174n48. See also *Hesperus* (ship)

British Indian Association (Transvaal), 99, 101

British Slavery Abolition Act (1833), 24–25. *See also* abolition of slavery

British subjects, 30, 91, 117; differentiation and, 93, 96, 110–11, 114, 117–19, 127–31, 136–37; legal rights of, 11, 19, 91–93, 119, 123, 126–27, 138, 188n12; liberty of, 31, 36, 50–52, 94, 126–27, 182n57; mobility rights and, 27–28, 113–16, 119, 131, 134; redefinition of, 11, 110, 126–27, 134–37; religion and, 94–96, 98, 100, 104, 108–9. *See also* liberalism; migration (free); race; *specific colonies*

Brubaker, Rogers, 6, 10

bureaucracy, 2, 13, 17, 55–67, 77, 80–83, 144–45, 148, 158n56. *See also* regulations

bureaucratic discretion, 65, 82, 90, 124, 128, 134, 142. *See also* administrative law

Butler, Judith, 168n105

Cachalia, A. M., 101

Calcutta, 25, 32–33, 57, 59–60, 68, 123, 131

Canada, 2, 8, 20–21; anti-Asiatic Leagues and, 121–22; citizenship and, 188n12; continuous journey regulation and, 123–28, 131–34; development of passport and, 112, 114, 116, 119–21, 125, 127, 131, 136–37; Indian immigrants and, 112, 116–20, 125–26, 130–33; *Komagata Maru* (ship) and, 132–35, 138, 193n80; labor demand in, 117–19; migration regulations and, 119–20, 123–26, 143; racial anxieties in, 112–17, 128–30, 134–39, 144

Canada Pacific Railway Company, 123

Canadian Immigration Act (1910), 132

Cape Colony, 87, 90, 97, 179n27

Cape Malays, 95

Caribbean, 16, 25, 53, 115. *See also* British Guiana; Jamaica; Trinidad

Carter, Marina, 164n63

certificate of residence. *See* Black Act (Transvaal, 1907)

Chakrabarty, Dipesh, 18

Chanock, Martin, 90, 93

Charles, James, 34

Chatterjee, Partha, 11, 146–47, 196n11

Chicago School, 4

children and legal status, 183n65

China, 120, 128–29, 132

Chinese indentured labor, 179n21, 181n37

Chinese subjects (in British colonies), 91–92, 110, 120, 128–30, 181n37

cholera, 174n48. *See also* health regulations

Christianity, 20, 108–9. *See also* marriage; religion

citizenship rights, 11, 19, 91–93, 119, 123, 126–34, 138, 146–50, 155n35, 188n12

civilizing mission (of colonialism), 8, 23, 30, 48–49, 51, 122, 139

civil law, 93

Clark, W. H., 131–32

class, 88–89, 95, 104, 115

climate, 112, 117, 119, 121, 124–25, 173n43

Clinton, Hillary, 142

Cohn, Bernard, 122

colonialism: anti-colonial movements and, 67, 85, 104–5, 120–21, 136, 144; the civilizing mission of, 8, 23, 30, 48–49, 51, 122, 139; expansion of, 22; features of, 11–12; indirect rule and, 93, 96, 171n19; nation-state formation and, 1–5, 8, 19–21, 85–87, 103–5, 107–14, 137–39, 144–50; rule of colonial difference and, 11, 29–30, 54, 114, 119, 136, 139, 143, 146; Westphalian ideal and, 7, 137–38, 154n23, 155n26. *See also* British subjects; race; racism

Colonial Office, London, 21, 25, 51, 94–95, 119, 128

colonial rule, 11, 49, 96, 120, 146, 150. *See also* British subjects; colonialism

colonial state: differentiated legal regimes in, 11, 21, 90–93, 135–37, 146–50; disciplinary power and, 17–18, 62–84, 144–45; migration regulation and, 54–62; and relation to modern state, 1–4, 11–17, 21, 24, 32, 42, 55, 84, 127, 138, 141–50. *See also* empire-state; liberalism; nation-state

Comoroff, John, 158n60

comparative advantage, 45

consent: contract and, 23–24, 33–43, 56, 61, 63, 114, 160n5; freedom and, 9, 16, 28, 44, 48, 52–54, 74, 143

constraint (logic of), 2, 9, 19, 55, 135

continuous journey regulation (Canada), 123–28, 131–34

contract: consent and, 16, 23–24, 36–42, 44, 50, 52–54, 143, 160n5; deviations from the norm and, 160n7; Dickens

Committee Report and, 34–41; equality in exchange and, 23, 41–42, 52–54; fraud and, 23, 35–38, 40–41, 49, 54, 56, 62–64, 177n91; freedom and, 29, 34–35, 38–40, 44; indenture and, 2, 16, 23–24, 36–44, 50–54, 143, 160n5; Maine on, 60–61; Marx on, 43–44, 48; performative and, 48, 168n105; state-authorized, 2, 8, 16–17, 31–32; validity of, 23, 25, 41, 165n77, 166n80. *See also* migration (indentured)

Coolies, 35–36, 52, 67, 71–73, 79

coproduction, 9, 12, 147–48

Council of India, 60

Court of Directors, 1, 26, 31, 67. *See also* East India Company

Craven, Paul, 24, 42, 166n81

Crosby, Alfred, 156n39

customary law, 93, 109, 181n40. *See also* liberalism

Daru, N. D., 118, 121

Dealers' Licenses Act (Natal, 1897), 90. *See also* franchise laws

decolonization, 118

Defence of India Act (1915), 137

D'Epinay, P., 29–32, 36

Derrida, Jacques, 168n105

DeWind, Josh, 154n21

Dick, G. F., 27

Dickens, Theodore, 33

Dickens Committee Report, 33–41, 45–49, 56–57

dietary scales, 18, 60, 76–77, 82–83. *See also* regulations

differential racism, 118. *See also* racism

differentiation: by gender, 19–20, 85–88, 98–106, 110, 144; of legal regimes, 11, 21, 109–14, 118, 128–33, 136, 149–50, 159n66; of mobility, 21, 119, 128–33, 139; by race, 19–21, 86–95, 104–10, 116, 130–36, 155n30

diffusionist model (of the modern state), 3, 12, 24, 42, 114, 127, 145–46

disciplinary power, 17–18, 30–31, 48,

58, 62–84, 158n60. *See also* sovereign power

discrimination, 87–93, 116, 121, 126, 128. *See also* racism

disease. *See* health regulations

dispersal model (of state formation). *See* diffusionist model (of the modern state)

District Magistrate, 63–64

Dowson, William, 34–38, 46–47, 51, 166n109

Dua, Ena, 192n65

Dutt, Russomoy, 34

East India Company, 1, 26, 31, 151n3

Eltis, David, 9, 155n30

embarkation depots, 68–69, 71

emigrant (legal definition of), 115–16, 119, 188nn9–10

emigration: barring of, 153n20; to Canada, 117–21, 125, 131, 134–35; labor contract and, 26, 45–46, 51, 56–58; to Mauritius, 45–46, 51, 53–57; regulations on, 56–71, 77–83, 114–16; to South Africa, 88–94, 101–6, 116, 120–21, 178n16; state control of, 7, 17, 55–60, 62–82; theoretical importance of, 4, 7. *See also* migration (free); migration (indentured)

Emigration Acts (India): Act V (1837), 26; Act VII (1871), 71, 78; Act XIII (1864), 61, 63, 65, 68, 82; Act XIV (1839), 33; Act XV (1842), 57–58, 63, 78, 81–82; Act XXI (1843), 57; Act XXI (1883), 78, 115, 119–20; Act XXV (1845), 58; consonance of, 59. *See also* legislation

Emigration Agent, 63–64, 71, 74, 79. *See also* Protector of Emigrants

Emmer, P. C., 164n63

empire-state: discrimination and, 91, 127, 129, 134; liberalism and, 91, 105, 126, 134; migration control and, 8, 28, 105, 116, 139; nation-state and, 1, 4, 85–86, 96, 110, 143, 151n1; religion and, 108–9. *See also* colonial state; modern state

Enthoven, R. E., 136, 138

equality in exchange, 23, 41–42, 52–54.
    See also contract
Esop, Hassan, 97
Esop v. the Minister of the Interior, 97
eventful nationalism, 6, 86, 114, 138. *See
    also* gender; marriage; nationalism

facilitation (logic of), 2, 19, 55, 85, 135
Federation of International Lacrosse,
    141–42
Fiji, 2, 14, 67, 115, 158n58, 176n83
Fitzgerald, David, 153n14
Fletcher, Ian, 136
Foucault, Michel, 13–14, 17–18, 55, 58, 62,
    79, 122, 144
franchise laws, 90, 179n18, 185n93
fraud (legal category), 23, 35–38, 40–41,
    49, 54, 56, 62–64, 177n91. *See also* con-
    tract; regulations
freedom, 16, 19, 23, 38–39, 48, 168n105. *See
    also* abolition of slavery; contract; lib-
    eralism; migration (indentured)
free labor contract, 17, 24, 32, 39, 42, 143.
    See also consent; contract
free migration. *See* migration (free)
free movement, principle of, 14, 27–29,
    55, 115, 135, 138, 143. *See also* migration
    (free)
free passengers (South Africa), 89–90, 95.
    See also migration (free)
French Revolution, 27, 145

Gandhi, Kasturba, 99
Gandhi, Mohandas Karamchand:
    London All-India Moslem League
    and, 92–102; South Africa and, 19–20,
    86–87, 106, 144, 179n27, 180n36, 183n65,
    183n68
gender: Black Act (Transvaal, 1907) and,
    92; male-female ratios and, 88; migra-
    tion and, 6, 19–20, 70, 85–87, 93–98,
    110; nationalism and, 5–6, 19–20, 85–87,
    98–106, 109–10; race and, 87–93, 95,
    104, 106, 109–10, 128; satyagraha and,

98–106, 144. *See also* eventful national-
    ism; marriage; wives; women
Geoghegan, J., 60–61, 81, 163n48, 171n15,
    173n38
German Overseas Empire, 147
Gillanders, Arbuthnot & Co., 25
Gillian, R. W., 135–36, 138
Gilroy, Paul, 113, 117–18
girmit, 16
girmitya, 16
Gladstone, John, 25, 32, 105
Gladstone, William, 25
Gladstone Coolies, 25, 33. *See also* British
    Guiana
Glenelg, Baron, 32–33
Glick Schiller, Nina, 152n12
globalization, 139, 195n97
Gokhale, G. K., 104
Gordley, James, 23, 41, 165n77
Governor General of Canada, 117, 119,
    126, 128
Grant, James, 34, 46–48, 50, 52, 56–57
the great experiment (Emancipation), 25.
    See also abolition of slavery; contract
Grey, Albert, 119
Griffiths, Hollier, 27–29, 31–32, 36, 50, 54
Guyana. *See* British Guiana

Habib, Seth Haji, 180n36
Hamidia Islamic Society, 103
Harcourt, Lewis, 108
Hardinge, Charles, 136, 138
Harris, Karen, 110
Haudenosounee, 141–42
Hay, Douglas, 24, 42, 166n81
health regulations, 72–78, 175n59
Hesperus (ship), 25, 161n15, 174n48
Hindus, 97–98, 107, 119, 121. *See also* India;
    South Africa
Hobbes, Thomas, 43, 52
Hollifield, James, 154n21
Hong Kong, 115, 122–23, 132
Hopkinson, William Charles, 130–32
House of Lords, 33

Hugon, Thomas, 51
Huttenback, Robert, 178n16, 179n29, 185n91

immigration: assimilationist paradigm and, 4; Canadian legislation on, 119–20, 123–24, 132; discrimination and, 116–17, 121–22, 125, 135; health regulations and, 72–78, 175n59; national identity and, 85–87, 96, 103–4, 109–10, 113, 118–20; South African legislation on, 87–98, 101, 106; sovereignty and, 24–33, 36, 52–55, 84, 112–14, 127, 133–47; Stanley on, 53; state control of, 7, 57–84, 86–93, 123–33, 138, 181n44; study of, 1, 3, 6–8, 13–21. *See also* emigrant (legal definition of); migration (free); migration (indentured)
Immigration Restriction Act (Transvaal, 1907), 93
Immigration Restriction Act (South Africa, 1913), 101
imperial liberalism. *See* liberalism
indentured labor, 2–3, 8, 33, 38–40, 46–48, 52, 54, 88–90, 137, 166n109. *See also* abolition of slavery; migration (indentured); regulations; *specific colonies*
indentured migration. *See* migration (indentured)
India: anti-colonialism in, 120–21; colonial rule and, 16, 21, 23–26, 31, 48–61, 91–92, 115–16, 120–22, 124, 126–28, 131–32, 135, 137; labor force of, 1, 25; Law Commission of, 26, 31; nationalism and, 19, 85–87, 98. *See also* Canada; colonialism; migration (free); migration (indentured); South Africa
Indian National Congress, 105, 120
Indian Opinion, 94, 98–103
Indian Political Association (Kimberly), 102
the Indian Question, 93, 98
Indian Relief Bill (South Africa, 1914), 106, 109–10

Indian Revolt of 1857, 67, 175n66, 179n28
indirect rule, 93, 96, 171n19. *See also* customary law; liberalism
industrial residence, 60, 88, 168n104, 171n14
inheritance, 99, 107, 183n65
investigative modalities, 122
Iroquois, 141–42
Islam, 95, 102
Ismail, Adam, 93

Jacobins, 27–28
Jamaica, 58–59
Japan, 120
Japanese subjects, 128–29
Jordan, H. H., 96–97, 100
juridical labor contract. *See* contract
Jussat, Ebrahim Mahomed, 94
Jussat, Fatima, 94, 97

Kale, Madhavi, 160n10, 163n48, 164n59
Khan, Mahboob, 102
King, Mackenzie, 122
kinship, 20, 86–87, 109–10. *See also* gender; marriage
Komagata Maru (ship), 132–35, 138, 193n80. *See also* Canada
Kumar, Ashutosh, 164n63

labor: abolition of slavery and, 16, 19, 22–25, 42–48, 50–54, 160n7; Canada and, 117–19; as a commodity, 43–47; contract and, 8, 23–24, 39–41, 48, 114; definition of, 115; demand for, 1, 86, 160n10; free, 22, 24–25, 45, 50; indentured, 2–3, 8, 33, 38–40, 46–48, 52, 54, 88–90, 137, 166n109; strikes, 105–6; wages of, 45–47, 50, 169n109. *See also* migration (free); migration (indentured)
labor power, 43–44, 48
lacrosse, 141
Lal, Brij, 164n63, 176n83
land rights, 88–91

dom and, 14, 29, 55, 115, 135, 143; gender and, 19–20, 85–87; kinship and, 19–20; logic of constraint and, 2, 9, 19, 135; logic of facilitation and, 2, 19, 85, 135; logic of restriction and, 86, 112; marriage and, 19–20, 87–88, 93–106; nationalism and, 3–6, 85–87, 103, 109–10, 118, 120, 144, 187n120; passports and, 112–14, 116, 119–21, 125, 127, 131, 136–37, 145, 188n11; race and, 19, 21, 86–87; South Africa and, 89–90, 93–95, 104–5; state control over, 1–10, 13–19, 26, 28, 113–14, 136, 138, 141–43, 145; study of, 9–10, 13, 15, 29. *See also* emigration; immigration

migration (indentured): British Guiana and, 25, 32–33, 54, 58–59, 69, 81, 174n48; bureaucracy of, 2, 13, 17, 55–67, 77, 80–83, 144–45, 148, 158n56; contracts and, 2, 16, 23–24, 34, 36–42, 44–45, 50, 52–54, 143, 160n5; Dickens Committee Report and, 33–41, 45–49, 51, 56–57; embarkation depots and, 68–69; fraud and, 35–38, 40, 49, 54, 56, 62–64, 177n91; logic of constraint and, 55; logic of facilitation and, 55; Mauritius and, 2, 25–31, 39–40, 45–46, 51, 54, 57, 59–60, 69; mortality rates and, 67, 76; principle of free movement and, 14, 27–29, 55, 115, 135, 143; recruiters and, 35, 54, 63–64, 172n29; regulations on, 57–84; rule of colonial difference and, 29–30, 54, 114, 139, 143; ships and, 68–70; South Africa and, 88–89, 101, 105; state control of, 8, 14, 16, 26, 49–61, 85, 114–15, 135; suspension of, 54, 89, 176n83, 177n91, 178n16. *See also* abolition of slavery

modern state: citizen/migrant distinction in, 21, 148–50, 155n35; colonial genealogy of, 1–4, 11–17, 24, 32, 42, 55, 84, 127, 138, 141–50, 157n51; diffusionist model of, 3, 12, 24, 114, 127, 145–46; formation of, 1–5, 8, 10–15, 17–19, 42, 55, 71–72, 80–92, 109–14, 127, 137–39, 144–50;

migration control and, 13–14, 24, 55–61, 84, 141–45; racism and, 19, 21, 86, 89, 95, 104, 112–13, 116, 122, 133–34, 139, 155n30; sovereignty and, 1, 3, 6–8, 13–21, 24–33, 36, 52–55, 84, 112–14, 127, 133–47. *See also* empire-state; nation-state

Mohapatra, Prabhu, 174n55

Moji, Japan, 132

Mombassa, 115

monogamous marriage, 20, 86, 97, 101, 104, 108, 159n66

Morley-Minto reforms, 190n30

Mouat, F. J., 76–77, 175n66

Muslims, 88, 97–98, 107

Natal, 88–91, 102, 106, 109, 180n30

Natal Government Railway, 88

Natal Indian Congress, 185n93

Natal Supreme Court, 102

nationalism, 2, 85–87, 96, 103, 118, 120–21, 187n120; developmental and eventful approaches to, 6, 86, 136; gender and, 5–6, 19–20, 85–87, 98–106, 109–10. *See also* satyagraha; South Africa

nationality, 6, 19–21, 109–13, 127, 130, 136, 139. *See also* passports

nationalization, 110, 118, 148, 154n22, 159n65

"The Nation Form" (Balibar), 112

nation-state: empire-state and, 1, 4, 85–86, 96, 110, 143, 151n1; formation of, 127, 137; methodological nationalism and, 4; migration control and, 21, 113, 139, 148, 150; sovereignty and, 21, 127, 133, 137. *See also* colonial state; modern state

Nehru, Jawaharlal, 105

(neo)Eurocentrism, 10, 145–46, 156n39

Newcastle (South Africa), 105–6

Nicolay, Sir William, 32

Niobe (ship), 193n80

No One Is Illegal, 149

Northrup, David, 38

"A Note on Emigration from India" (Geoghegan), 60

Oliver, Frank, 133, 135, 138
Orange Free State, 87, 90
Orientals, 127–28. *See also* Anti-Asiatic
Leagues

Panama Maru (ship), 131–32
Pankhurst, Emmeline, 100
Parsis, 88
Partridge, G. B., 68–69, 74–75, 80
passive resistance. *See* satyagraha
passports, 2, 20–21; development of,
112–14, 136–37, 188n11; Canada and,
112, 116, 119–21, 125–27, 131; Haudeno-
sounee/Iroquois nation and, 141–42;
Komagata Maru (ship) and, 132–35, 138;
national identity and, 113, 136–39, 143;
state sovereignty and, 113–14, 134–139,
141–43; United States and, 142. *See also*
race; regulations
performative, the, 48, 168n105
Pietermaritzburg, 102
Plantation Manager, 66–67
plantations: labor demand in, 1–2, 16–17,
49–51, 143, 160n10; as sites of labor control, 65–71, 80–81, 88, 143, 148, 172n29,
176n83; strikes and, 105–6
*Political Theory of Possessive Individual-
ism, The* (Macpherson), 43
polygamous marriage, 19, 95–98, 101, 106,
108–9, 182n58, 184n85
Port Elizabeth British Indian Association,
102
Portes, Alejandro, 154n21
possessive individualism, 43–44
post-national globalization, 139
Prakash, Gyan, 157n51
principle of free movement, 14, 27–29,
55, 115, 135, 138, 143. *See also* migration
(free)
principle of reciprocity, 8, 136–37. *See also*
passports
property rights, 89–91, 96
Protector of Emigrants, 56, 62–65, 71
purna swaraj, 105

race: climate and, 112, 117, 119, 121, 124–25;
gender and, 87–93, 95, 104, 106, 109–10,
128; law and, 90–93, 106–12, 130, 134,
136, 166n81, 179n20, 181n40; migration
control and, 19, 21, 86–91, 116, 139,
155n30; passports and, 112–13, 116–21,
125, 127, 132–35, 137–38, 141–42; racial
superiority and, 89, 95, 104, 133; rule of
colonial difference and, 11, 29–30, 54,
114, 119, 136, 139, 143, 146. *See also* British subjects
racism: Canada and, 112–17, 121, 130,
133–39, 144; differential, 118; immigration regulations and, 122, 124, 128–37;
nationality and, 19, 21, 86, 89, 95, 104,
116, 133, 136, 139, 155n30; passports and,
132–35, 137–38, 141–43. *See also* bureaucratic discretion; rule of colonial
difference
Rainbow (ship), 193n80
Rasul, Bai, 93, 97
reciprocity, principle of, 8, 136–37
Reciprocity Resolution (India, 1917), 136
record keeping, 78–80
recruiters, 35, 54, 63–64, 172n29
regulations: annual reports as, 64–65; on
Canadian migration, 123–26, 132–33;
continuous journey, 123–28, 131–34;
disciplinary power and, 83–84, 144–45;
exceptions to, 82; exploitation and,
80–82; health, 72–78, 175n59; inconsistencies of, 59–61; on indentured
migration, 57–84; key figures of, 66–67;
and marriage, 87; for migration control, 1–3, 9, 13–14; multiplication of,
82; race-based, 19, 21, 86–93, 106–16,
130–39, 155n30; record keeping and,
78–80; sites of, 67–78; South African
provinces and, 93, 181n44; sovereignty
and, 1, 3, 6–8, 13–21, 24–33, 36, 52–55,
84, 112–14, 127, 133–47; surveillance
and, 62–65; taxes and, 89–92, 101–6,
116, 120, 178n13
Reid, Malcolm, 193n80

religion, 20, 94–96, 98, 100, 104, 108–9. *See also specific religions*

resistance. *See* revolutionaries; satyagraha

restriction (logic of), 55, 86, 112

revolutionaries, 130–32

Rex v. Sukina, 96

Richmond, Theophilus, 161n15, 174n48

riots, 121

rule of colonial difference, 11, 29–30, 54, 114, 119, 136, 139, 143–47, 196n11. *See also* liberalism

Russell, John, 51, 163n48

Sarah (ship), 25

Sartori, Andrew, 159n3

Saskatchewan, 127–28

satyagraha, 19–20, 86–87, 89, 92–93, 98–108, 144, 183n68, 185n91. *See also* Gandhi, Mohandas Karamchand; gender

Schuler, Monica, 161n11

Sea Lion (ship), 193n80

Searle, Malcolm, 97–104, 107, 109

sexuality, 2, 6, 20, 85–86, 95, 107, 109–10, 159n66

Shanghai, 132

Sharma, Nandita, 156n36, 156n38

Shepstone, Theophilus, 180n30

Sikhs, 119, 125, 132

Singh, Sardar Gurdit, 193n80

Singha, Radhika, 173n33, 180n36

Sinha, Mrinalini, 159n65

Sirdars, 66, 172n29

Slater, S. H., 131–32, 135

slavery. *See* abolition of slavery

Smuts, Jan, 99

Solomon Commission, 97, 104, 106, 108–9, 183n65

South Africa: colonies of, 86–87, 178n5; Immigration Restriction Act (1913), 101; and Law 3 (Transvaal, 1885), 91–92; Supreme Court of, 94; taxes on ex-indentured Indians, 89–90, 92, 101–3, 105–6, 116, 178n13; treatment of Indian immigrants in, 88–92, 120–21. *See also* marriage; migration (free); satyagraha

South Africa British Indian Committee, 95, 102

South African war, 19, 87, 91–92

sovereign power, 17–18, 27, 50, 58, 83. *See also* disciplinary power

sovereignty. *See* sovereign power; state sovereignty

Stanley, Edward, 46, 52–54, 57–58, 167n96

state sovereignty: migration control and, 1–3, 9, 13–16, 20–33, 54–55, 135; nationalization of, 6–8, 84, 113–14, 135–39, 143; Westphalian system and, 7, 8, 127, 137–38, 153n21, 155n26

Steinmetz, George, 146–47

strikes, 105–6

Sturman, Rachel, 174n56

sugarcane, 51, 53

Supreme Court of South Africa (Transvaal division), 94

swaraj, 105, 120

Swayne, Eric, 119, 125

Tata, Ratan, 104

taxes, 89–90, 92, 101–3, 105–6, 116, 120, 178n13

Thistlethwaite, Frank, 156n37

Tinker, Hugh, 38–39, 163n48, 164n63

Torpey, John, 13, 145–46, 196n8

traders, 88–90

trading licenses, 90, 96, 185n93

transnationalism (in study of migration), 90, 96, 185n93

Transvaal, 87, 89, 91–97, 107

Transvaal Black Act. *See* Black Act (Transvaal, 1907)

Transvaal Indian Women's Association, 99, 104

Transvaal Leader, 94

Treaty of Westphalia, 7–8, 154n21, 155n26. *See also* Westphalian system

Trinidad, 58–59, 69, 76, 152n9, 173n41,
178nn9–10
Trump, Donald J., 191n44

Union of South Africa, 19–20, 87–92. *See
also* South Africa
United Kingdom, 141–42, 196n6
United States of America, 7, 116, 123, 141–42

Vancouver, 117, 121, 130, 132, 193n80
Victoria, British Columbia, 131–32

Waldinger, Ralph, 153n14
Warren, Falk, 118, 121
Wellesley (ship), 76
Wessels, John, 94–98, 100–101
West Indies, 60. *See also* Caribbean
Westphalian system, 7, 127, 137, 154n23,
155n26. *See also* Treaty of Westphalia

Whitby (ship), 25
white-settler colonies, 2, 9, 19, 29, 84,
90, 111, 116, 120, 143. *See also* race;
racism
white supremacy, 89, 104. *See also* race;
racism
will. *See* consent
Williams, Eric, 159n2, 169n113
wives, 87, 89, 93–109, 184n80. *See also*
gender; marriage
women, 19, 87, 89, 92–109, 128, 184n80.
See also gender; marriage
Wong, Diana, 10, 156n37
World War I, 137, 142
Wright, Barry, 160n5

Yokohama, Japan, 132
Young, Robert, 49